30-SECOND
GREAT INVENTIONS

30-SECOND
GREAT INVENTIONS

50 light-bulb moments that
changed the world, from the
compass to the smartphone,
each explained in half a minute

Editor
David Boyle

Contributors
David Boyle
Judith Hodge
Diana Rawlinson
Andrew Simms

Illustrations
Steve Rawlings

IVY PRESS

Leabharlanna Poibli Chathair Bhaile Átha Cliath

Dublin City Public Libraries

First published in the UK in 2018 by
Ivy Press
An imprint of The Quarto Group
The Old Brewery, 6 Blundell Street
London N7 9BH, United Kingdom
T (0)20 7700 6700 **F** (0)20 7700 8066
www.QuartoKnows.com

British Library Cataloguing-in-
Publication Data
A catalogue record for this
book is available from the
British Library

ISBN: 978-1-78240-512-2

This book was conceived,
designed and produced by
Ivy Press
58 West Street, Brighton BN1 2RA, UK

Publisher Susan Kelly
Creative Director Michael Whitehead
Editorial Director Tom Kitch
Art Director James Lawrence
Project Editor Stephanie Evans
Designer Ginny Zeal
Assistant Editor Jenny Campbell

Printed in China

10 9 8 7 6 5 4 3 2 1

CONTENTS

INTRODUCTION
David Boyle

What makes a successful inventor? There are many ingredients. You need to have the background knowledge in your field – like the Wright Brothers, American pioneers of the aeroplane. You need to know the right people who can help you exploit your invention, as Scottish engineer James Watt did when he famously got together with Birmingham manufacturer Matthew Boulton to make the most of Watt's improvements to the steam engine. Then you need people to understand the significance of what you have created; without this an invention can lie ignored and unexploited for decades, as Scottish inventor James Blyth found with his wind energy machine. You also need to have investors and you need to be able to protect your invention, as the rival inventors of the pneumatic tyre, Robert Thomson and John Boyd Dunlop, discovered. But most of all, you need a moment of inspiration.

Sometimes this is a revelation that is then worked out carefully in theory, as with Archimedes' screw, an ancient Greek invention. Sometimes it is a series of exhausting experiments to see how the theory works in practice, which might take years, as English engineer Frank Whittle found developing his jet engine. Or it appears, as the pioneering English economist John Maynard Keynes described, like a 'grey woolly monster in my head'.

When you look at a range of different inventions and inventors, as you can do side by side in this book, it becomes clear that to be successful inventions also have to come along at the right moment. Of course, as with any rule, there are exceptions to this. The mechanical clock, invented by the Chinese in the eighth century, failed to spread to the rest of the world until the twelfth century; the use of paper money, pioneered by Mongol ruler Kublai Khan in the twelfth century, did not reach Europe until the end of the seventeenth century. But generally speaking, successful inventions hit the spot because they are developed at a time when there is both the knowledge to make them possible and a need they can meet. This was true of Watt's steam engine – and also of the steam boat, plastic and the internal combustion engine.

This raises an important question. Were the inventors of these things the only people who could have made the breakthrough they did? Or did they happen to be in the right place at the right time so they could steal a few years or days on their competitors? If Watt had died young or become caught up in the aftermath of the efforts of Charles Edward Stuart ('Bonnie Prince Charlie') to regain the throne in 1745, and never ended up as a student in Glasgow, nor met his friend the physicist John Robison, or the other doyens of the Scottish Enlightenment, would somebody else have made the breakthrough he did with steam power?

It is impossible to be sure. But let's keep a place for the role of individual genius. The inventors in this book are all people who managed to understand their field so completely that they were able to see what else, which new development was possible. Yet at the same time, they brought some insight from outside the field to bear on the problem. American inventor Thomas Edison famously claimed that invention was 1 per cent inspiration and 99 per cent perspiration, and he went on to prove it in his tireless search for the right material to maximize the brightness and longevity of the filament in his light bulb. He had more than 1,000 patents to his name, so he may have proved this point – but there are other examples to the contrary. Nor do many people really believe that they might be an Edison if they just put in enough perspiration. No, moments of inspiration are vital, too: progress is underpinned, whether it is in technology, economics or social development, by moments of personal imagination.

In addition to all this inventions need wider networks of people to make the breakthrough idea a reality. They need connections with people powerful or wealthy enough to make change happen. This may explain why so many inventors are white men, because those critical networks may not have been available to anyone else. *30-Second Great Inventions* includes two brilliant twentieth-century women inventors. There must be many more whose graves are not marked and whose memories are not celebrated.

From the mind of a genius
Some great inventions, such as the X-ray, came about by chance, but their inventors needed the brains and insight to realize their life-changing significance.

How this book works

There are seven sections, designed to include as much of the science, technology and economics of inventions as possible. The first section looks at **Materials**, from cement through to petroleum and plastic. **Construction & Engineering** investigates key mechanical breakthroughs such as the mechanized clock, the internal combustion engine and the plough. In **Transport & Location**, we cover inventions connected to movement, from the pneumatic tyre to GPS satellite positioning. **Medicine & Health** probes the medical field, from antibiotics to pacemakers. **Communications** runs the gamut from the printing press to the smartphone – including, of course, Tim Berners-Lee and the Internet. **Economics & Energy** looks at everything from credit cards to solar energy. Finally, **Daily Life** highlights those inventions that many people now take for granted, from heating and lighting our homes to preserving food by canning and refrigeration.

Each entry in this book is accompanied by a pithy 3-second survey, which boils the facts down to the absolute basics. There is also a 3-minute overview, which tries to give a different and more challenging edge to the story, together with biographical facts for the key players.

This is not a book that has to be read from cover to cover. You can dip in here and there to check facts, follow the narrative of an individual's struggle against the odds or even derive inspiration from a great inventor's flash of insight that changed the world.

MATERIALS

MATERIALS
GLOSSARY

alloy Mixture of two elements, which sometimes has characteristics shared by neither of its component elements.

ancient Greeks Civilization of competing and cooperating city-states in Greece, dedicated to learning and sporting prowess, which thrived for 1,000 years, c. 500 BCE–c. 500 CE.

Assyrian empire Sprawling ancient empire in the Middle East, based around the cities of Assur and Nineveh, which thrived 2300–600 BCE.

Babylonian empire Ancient empire based around the city of Babylon in what is now Iraq, which rose and fell c. 1800–c. 500 BCE.

Bakelite The first successfully and widely marketed form of plastic.

bitumen Thick, oily mixture used for fixing roofs and making roads.

concrete Mixture of cement, stones and other aggregates that hardens over time.

Crystal Palace Glass structure designed by greenhouse designer Joseph Paxton for the Great Exhibition in London in 1851 in what is now Kensington Gardens. The structure was moved south of the river to Upper Norwood, where it burned down in 1936, but not before it had inspired a new generation of modern architects.

Eddystone Lighthouse There have been four lighthouses on this spot over 14 kilometres (9 statute miles) south of Rame Head, Cornwall, south-west England, designed to warn shipping away from the Eddystone Rocks. The third one, built by John Smeaton in 1759, used a new concrete that could set underwater and pioneered the design of lighthouses everywhere.

electroplating Process that uses an electrical current to cover an object in a thin coating of metal.

eunuch Man who has been castrated, as was common for officials in the Chinese court for some centuries.

French Revolution Uprising against distant, aristocratic rule that exploded in Paris in 1789 and led directly to 'the Terror', when the revolution turned into bloody reprisals against the former ruling elite.

greenhouse effect Phenomenon whereby burning fossil fuels leads to a kind of hothouse in Earth's atmosphere that traps solar heat, raising the temperature, melting polar ice caps and disturbing the weather. It was first identified in 1824.

Hittites People living in what is now Turkey whose empire emerged in c. 1600 BCE.

Industrial Revolution Emergence of an industrial economy, fuelled largely by available coal and steam power, which relied primarily on Scottish know-how and English finance. It had no clear beginning and end, but arguably started around 1770, after Watt's steam engine made it possible, and settled down to more normal speeds of innovation around 1820.

iron ore Rocks in which iron deposits can be found and iron extracted.

natron Basis for an early form of soap, a mixture between soda ash (sodium carbonate decahydrate) and bicarbonate of soda.

nitrocellulose Highly inflammable compound originally known as gun cotton.

Pantheon Round Roman temple, completed by Emperor Hadrian in 126 CE, which still stands in Rome, Italy, together with its innovative concrete dome. It has been a Christian church since the seventh century.

papyrus Thick paper made from the leaves of the papyrus plant and used first by the ancient Egyptians.

paraffin Refined fuel or lubricant derived from oil. Also known as kerosene.

pig iron Impure brittle iron from the bottom of a blast furnace, so-called because the shape of the ingots resemble piglets.

polymer Molecule made from joining a number of small molecules together. It can be man-made or natural – plastic and rubber are both polymers.

radial tyre Tyre made from strips of rubber-coated plastic arranged at right angles to the direction of travel.

Sheffield steel Steel made in Sheffield, in northern England. The city came to specialize in its manufacture, largely because of its track record of making cutlery dating back to the thirteenth century.

vellum Parchment originally made from the skin of a calf.

CEMENT

the 30-second history

The story of cement stretches

back into prehistory, when the ancient Greeks, Babylonians and, before them, the Assyrians began to experiment with binders capable of hardening to hold other materials together in a building. The Assyrians used bitumen and the Egyptians a mixture of sand and gypsum to create mortar, which was often mixed with small stones. But it wasn't until the Romans came along that the idea was perfected into a useful form of concrete that could be shaped in ways their forefathers had never dreamed possible. Roman architect Marcus Vitruvius Pollio – who died in 15 CE – set out the basic ideas in his *Ten Books of Architecture* and wrote down the ingredients for *pozzolana*, a crushed volcanic ash that allowed cement to set underwater. Vitruvius' principles were put into practice spectacularly in the building of the Pantheon, beginning around 118 CE for the Roman emperor Hadrian, which included the largest unsupported dome in the ancient world and is still standing to this day. The vast system of Roman aqueducts used cement and concrete in ways that would have been impossible for previous civilizations. Perhaps surprisingly, one of the casualties of the fall of Rome in 476 – and forgotten until the Renaissance – was the knowledge of how to make effective concrete.

3-SECOND BIOGRAPHIES
MARCUS VITRUVIUS POLLIO
c. 80–70 BCE–C. 15 CE
Roman architect and engineer whose work De Architectura contains the earliest descriptions of hydraulic setting cement developed by the Greeks and Romans to build their harbours

JOHN SMEATON
1724–92
English civil engineer who developed modern cements by experimentation with strong, quick-drying mixes that could set underwater

JOSEPH ASPDIN
1778–1855
English cement manufacturer who patented the first version of what became known as Portland cement, due to its resemblance to Portland stone – the Victorian cladding of choice

30-SECOND TEXT
David Boyle

Cement may seem modern but it was in use in prehistoric times.

3-SECOND SURVEY
Cement and concrete – a mixture of cement, sand or gravel, and water – are the foundations, so to speak, of civilization, because they are the critical binders that hold buildings together.

3-MINUTE OVERVIEW
The secrets of cement were not rediscovered until Renaissance scholars began to unearth Classical texts. But, modern breakthroughs happened, thanks to – among others – engineer John Smeaton. He researched ways to develop a cement that would set firmly within 12 hours, and between high tides, so that he could build the third Eddystone Lighthouse in the English Channel. It was successfully completed in 1759.

GLASS

the 30-second history

The volcano known as Vesuvius famously erupted in 79 CE, covering the cities of Pompeii and Herculaneum in southern Italy in volcanic ash. When the site was rediscovered, archaeologists found that some of the wealthier households in both cities had glass windows, which kept the weather out and still let in light, although unfortunately were not able to keep out the burning ash. It was a sign that, although the Romans were not the first to develop glass, they certainly drove forward its manufacture. Glass had been developed in the civilizations of the Bronze Age, including those in Syria, China and India, but especially ancient Egypt. The pharaohs oversaw the development of glass and, well into the Roman period, the ancient city of Alexandria produced the best-quality glass in the world. When the Emperor Hadrian visited there in the second century CE, he was so impressed that he sent a translucent glass vase back to his friend Servianus in Rome. And if it wasn't the Egyptians, it was the Phoenicians, who, according to the Roman writer Pliny the Elder, claimed that they discovered the phenomenon when their merchants cooked on sand under Mount Carmel, with their pots set on top of their cargo of *natron* (a natural preservative), and found, when they finished heating their food, that glass had formed underneath.

RELATED TOPICS

See also
CEMENT
page 16

THE OPTICAL LENS
page 84

3-SECOND SURVEY

So simple a material that it can be made from sand alone, glass keeps out weather, can be used for drinking vessels and decorative objects, in lenses and for much else.

3-MINUTE OVERVIEW

Roman manufacturers developed clear window glass (although their early versions were tinted green) and were able to mass-produce glass cups and bottles. But as with so many of their technological innovations, Roman techniques died out in the Dark Ages and had to be rediscovered. It was Italian and Bohemian merchants who created and developed the glass that dominated life in the fourteenth century and afterwards.

3-SECOND BIOGRAPHIES

**GEORGE RAVENSCROFT
1632–83**
English glass manufacturer and exporter who first discovered that adding lead to glass made it clearer and easier to shape

**ROBERT LUCAS CHANCE
1782–1865**
English glass manufacturer who, aided by the French glass technician Georges Bontemps, perfected the techniques for making plate glass

**JOSEPH PAXTON
1803–65**
English gardener whose design for London's Crystal Palace in 1851 shaped the way we use glass in buildings today

30-SECOND TEXT

David Boyle

Glass illuminates our daily life, and enables scientists to see what lies beyond our world.

PAPER

the 30-second history

Cai Lun, an official and a senior eunuch in the court of the Chinese emperor He of Han and his successors, is credited with inventing paper and the papermaking process in around 105 CE. He was reputedly inspired by watching paper wasps make their nests out of dried vegetable fibres. Cai Lun was involved in bitter internal disputes and jockeying for power as he backed the interests of the empress Dou and her entourage. At court, he rose to be in charge of manufacturing weapons and, in that role, is credited with first creating paper, which he managed using tree bark, old rags and fishing nets. He subsequently drank poison when the new emperor ordered him to report to prison for backing the wrong side. But Cai Lun was revered in the next century when papermaking spread rapidly through China, accelerating the growth of the literary culture that helped the Chinese to develop politically and culturally. It was not for another 500 years, when papermakers were captured by Arabs in the seventh century, that papermaking techniques were taken back to the Arab world, and then via returning crusaders, eventually reached western Europe.

RELATED TOPIC
See also
THE PRINTING PRESS
page 96

3-SECOND BIOGRAPHY
CAI LUN
48–121 CE
Senior Chinese eunuch and professional inventor who is credited with inventing modern papermaking

30-SECOND TEXT
David Boyle

3-SECOND SURVEY
Before paper, anything written down had to be scrawled onto expensive papyrus or vellum. The invention of cheap paper meant that ideas could be disseminated more easily.

3-MINUTE OVERVIEW
By 751 CE, the Chinese Tang dynasty had spread as far as the edge of the Gobi Desert. That year the Battle of Talas River halted their expansion and handed control of the Silk Road to the Muslim world. It was then that captured Chinese papermakers were said to have taken their techniques to the Mediterranean, although it took another 500 years for them to reach western Europe.

The nests of the paper wasp, constructed from plant material, provided inspiration for the development of papermaking techniques in China in the first century.

31 July 1923
Born in New Kensington, near Pittsburgh in Pennsylvania

1933
Kwolek's father dies – she credited him with introducing her to the wonders of the natural world

1942
Begins her university studies at what is now Carnegie Mellon University, choosing chemistry because she can't afford the fees for medicine

1946
Offered a job with DuPont

April 1959
Her award-winning paper describing 'the Indian rope trick' is published, explaining a classroom experiment to make nylon in a jar

1964
Carries out the experiment that led to the discovery of Kevlar

1971
Kevlar is marketed by DuPont for the first time. It has since earned billions of dollars for the company

1986
Retires from Du Pont having led the research team on polymers for some years

1995
Awarded the DuPont company's Lavoisier Medal for outstanding technical achievement

June 2014
The millionth bulletproof vest using Kevlar is produced

18 June 2014
Dies aged 90

STEPHANIE KWOLEK

It was one of those moments of serendipity, and happened when industrial chemist Stephanie Kwolek was working as part of a team at chemical giant DuPont's research laboratory at Wilmington in the United States. It meant that she stumbled on a new kind of plastic fibre five times stronger than steel. The result was marketed as Kelvar, and was used all over the world – best known as the lightweight basis for bulletproof vests.

'I knew that I had made a discovery,' Kwolek said later. 'I didn't shout "Eureka", but I was very excited, as was the whole laboratory excited, and management was excited because we were looking for something new, something different, and this was it.'

She was supposed to have been looking for a light fibre that could replace the steel in radial tyres. That involved changing a series of carbon-based molecules into larger ones known as polymers. In 1964, she was trying to convert one of these into liquid form and it came out disappointingly muddy. But she carried on, persuaded her colleagues to risk it in the spinning machine and found that the liquid produced an extremely stiff fibre. Not only did it turn out to be far stronger than steel of equal weight, but it also resisted fire.

Kelvar began as 'Fibre B', a new ingredient for tyres. Aside from protective clothing, the fibre is now found in a whole range of products, including aeroplanes, mobile phones and sails.

The daughter of working-class Polish immigrants, Kwolek wanted to be a doctor but couldn't afford the tuition for medical school. She studied chemistry at what is now Carnegie Mellon University and looked for a temporary job in chemical research to fund her medical studies. Her future mentor Hale Charch – one of the pioneers who had just produced nylon – interviewed her and, hearing that she already had a job offer elsewhere, offered her a contract on the spot.

It was not until 1975 that Kevlar itself hit the market, and Kwolek's invention earned billions for DuPont. She made nothing out of it directly because she had signed over her patents to the company.

David Boyle

IRON

the 30-second history

About 5 per cent of Earth's crust is composed of iron ore – the rocks and minerals that contain iron in its natural state. 'Smelting' is the process of extracting the pure iron from its ore. Who first smelted iron? Nobody knows exactly, but the smelting process seems to have been invented towards the middle of the Bronze Age by the Hittites, a civilization in modern Turkey that reached its height in the fourteenth century BCE. When the Hittite empire fell, in c. 1180 BCE, their methods of smelting iron began to spread more widely to southern Africa and India, and then Greece and China. But the key to modern iron was the development of cheap production methods in the Industrial Revolution, including those for producing steel, which made possible so much of the rest of industrialization. One man in particular made this a reality: a Quaker farmer's son, Abraham Darby who, in 1709, invented a blast furnace that used coke rather than charcoal in order to reach temperatures suitable for producing cast iron in major quantities. By the time Abraham Darby's grandson took over the business, they had made enough to construct the very first large-scale human structure to be made entirely of iron – the iron bridge across the river Severn in Shropshire – which gave its name to the town where it was built.

RELATED TOPIC
See also
STEEL
page 24

3-SECOND SURVEY
Iron is one of the most common minerals on Earth. Techniques to extract it, purify it and use it on a large scale made possible the Industrial Revolution in the eighteenth century.

3-MINUTE OVERVIEW
During the reign of the fourteenth-century English monarch Edward III, iron was considered so valuable that a number of the iron kitchen pots in his household were categorized as jewellery. The medieval period produced new furnaces that allowed for new cast iron shapes to be produced, with the royal pots being prime examples.

3-SECOND BIOGRAPHIES
ABRAHAM DARBY
1678–1717
English Quaker pioneer of modern iron smelting

THOMAS FARNOLLS PRITCHARD
1723–77
English architect and interior designer, who designed the world's first all-iron bridge at what is now Ironbridge, a town in Shropshire

JAMES BEAUMONT NEILSON
1792–1865
Scottish inventor who developed the hot blasting process that increased the efficiency of iron smelting

30-SECOND TEXT
David Boyle

The process of smelting iron was mastered during the Bronze Age, but it required steam power to make iron smelting possible on an industrial scale.

STEEL

the 30-second history

The Tay Bridge Disaster of 1879 occurred when a wrought iron bridge over the river Tay in Scotland collapsed during a storm, plunging a packed passenger train into the water below. It was the last in a series of similar disasters involving wrought iron bridge collapses. Engineers knew that wrought iron was not the most suitable material for building bridges and other large-scale structures. They needed something purer, harder and stronger. But steel, which had been made as an iron alloy since at least 1800 BCE, led by the Tamils in India, was too expensive to make in large quantities. The man to change that was Henry Bessemer, whose father had worked on iron alloys at the Paris mint before fleeing to England during the French Revolution of 1789. Bessemer developed a method that involved blowing oxygen through molten pig iron to burn off impurities, which he unveiled in Cheltenham in 1856. In the audience was James Nasmyth, a Scottish engineer who had been working on the same problem; he abandoned his solution once he heard Bessemer was thinking along the same lines. Bessemer offered him a third of the patent but was turned down because Nasmyth wanted to retire. The city of Sheffield's link with steel was established when Bessemer launched a steel foundry there to produce steel using his new method.

RELATED TOPIC
See also
IRON
page 22

3-SECOND SURVEY
An iron-carbon alloy, steel may not be the strongest substance known to humankind but it is the toughest and hardest practical material that can be produced cost-effectively.

3-MINUTE OVERVIEW
In 326 BCE Porus, king of a region of India that is now the Punjab, presented Alexander the Great with a steel sword. The Tamils in Sri Lanka and southern India seem to have led the way with steelmaking in the ancient world – although the oldest archaeological evidence of steel production is actually in Anatolia (modern day Turkey) from c. 1800 BCE.

3-SECOND BIOGRAPHIES
HENRY BESSEMER
1813–98
English engineer, originator of the Bessemer process and many other inventions, including a ship that kept passengers level so they never suffered from seasickness

ANDREW CARNEGIE
1835–1919
Scots-American who made a fortune from steel production and became, at one time, the richest man in the world

30-SECOND TEXT
David Boyle

High tensile strength and low manufacturing costs enable steel to be widely used in major construction projects, from skyscrapers to suspension bridges, railways to rivets.

PETROLEUM

the 30-second history

In 1847, Scottish chemist James Young noticed that black oil was seeping out of the roof of a colliery in Derbyshire. He began experimenting with the oil and managed to produce a lighter fuel oil for lamps and a heavier oil for lubricating machinery. Four years later, the mine's supply of oil had begun to dry up, but Young discovered it had been seeping through the sandstone roof of the mine and realized that it might be possible to create it artificially. It was this idea that allowed him to produce what he called 'paraffine oil' or paraffin. The invention made his fortune, eventually leading to the mining of shale oil deposits in Scotland to produce it. Before his great breakthrough, Young had educated himself at night school in Glasgow and became a close friend of the missionary David Livingstone. After his discovery, Young spent his money on yachting, scientific experiments and in funding Livingstone's African explorations. The word 'petroleum' derives from the Latin word for rock, and rock oil was the phrase used to describe the substance that comes out of the ground, then mainly from Baku in Russia, where the first oil refinery was built in 1861.

RELATED TOPIC
See also
INTERNAL COMBUSTION
ENGINE
page 48

3-SECOND SURVEY
Petroleum products now dominate the world economy, not always to the advantage of the nations that supply the oil. Economists call the side effects of these bonanzas the 'Curse of Oil'.

3-MINUTE OVERVIEW
The realization that oil could be refined into fuel and other necessities turbo-charged twentieth-century economics, but it has also contributed to raising the overall temperature of the planet, changing the climate and boosting the so-called 'greenhouse effect' – a phrase coined in 1917 by Scottish inventor Alexander Graham Bell.

3-SECOND BIOGRAPHIES
JAMES YOUNG
1811–83
Scottish chemist, inventor and philanthropist who first made paraffin and is credited with the breakthroughs that allowed petroleum to be refined

JAMES MILLER WILLIAMS
1818–90
Canadian businessman, railway manufacturer and politician who launched the petroleum industry in Canada, and pioneered similar businesses across North America

JAN JÓZEF IGNACY
ŁUKASIEWICZ
1822–82
Polish pharmacist who built the first working oil refinery and invented the kerosene lamp

30-SECOND TEXT
David Boyle

Mining oil transforms landscapes – just as it has transformed economic fortunes.

PLACTICS

the 30-second history

Metallurgist Alexander Parkes

was an expert in electroplating before he was credited with inventing the first man-made plastic material, which he called Parkesine. Patented in 1862, his invention treated cellulose with nitric acid and a solvent to create what was actually nitrocellulose – now popularly known as synthetic ivory. He went into business making his material in a factory in Hackney, in east London, but the venture failed because Parkesine was highly flammable, cracked easily and was rather expensive to produce. Some of Parkes's 17 children helped in his experiments, which tried to replicate some of the properties of natural materials like rubber or ivory. But Parkesine was less successful than its later rivals, xylonite and celluloid, which took its place in the latter years of the nineteeth century. All three were then overtaken by the arrival of Bakelite in 1907. The invention of Belgian-American chemist Leo Baekeland, Bakelite was the first completely synthetic plastic material and appeared just in time for it to be used widely during the First World War. Bakelite's full chemical name was polyoxybenzylmethylenglycolanhydride, and it had magical insulating properties that meant it was unable to conduct heat or electricity. This made Bakelite perfect as a material for telephones and other electrical gadgets.

RELATED TOPIC
See also
THE CREDIT CARD
page 130

3-SECOND SURVEY
Before the invention of plastics nearly all household objects were made from natural products such as bone or wood; plastics are much harder wearing and can be mass-produced cheaply.

3-MINUTE OVERVIEW
Bakelite inventor Leo Baekeland faced legal action to defend his patents in the United States, and winning his case led to a merger between three companies – his own included – in 1922 to form the Bakelite Corporation. The trademark they used incorporated the mathematical symbol for infinity, and they sold their product aggressively under the slogan: 'The material with a thousand uses.'

3-SECOND BIOGRAPHIES
ALEXANDER PARKES
1813–90
English metallurgist who invented Parkesine

LEO BAEKELAND
1863–1944
Belgian-American chemist who began experiments after selling his photographic paper patents to Eastman Kodak and promising not to work on photographic products for 20 years. The result of his experimentation was Bakelite

30-SECOND TEXT
David Boyle

The Bakelite telephone was one of the first plastics found in people's homes.

THE WHEEL

the 30-second history

3-SECOND SURVEY
Wheels sprung up simultaneously in many parts of the world, enabling people to move heavy objects vast distances.

3-MINUTE OVERVIEW
Although several cultures could claim responsibility for inventing the wheel, there are some that clearly can't. The ancient American civilizations did come up with a wheel, but they used it as a children's toy. The difference was that, unlike in Asia and Europe, the original Americans had no horses, and buffalo did not take kindly to domestication and pulling carts.

Who was it who first invented the wheel? This is important because it is, in some ways, the most significant discovery ever made by human beings. The trouble is that the answer is lost in the mists of time. You might as well ask which human being first stood upright on two legs. Not even the experts agree, but there is a prevailing opinion that the first wheels were actually potter's wheels, in Mesopotamia, sometime around 3500 BCE. The earliest example to be found by archaeologists was unearthed at the ancient city of Ur, in what is now Iraq. Carbon dating suggests that it was made around 3129 BCE. But it is confusing that, in fact, wheels appeared at the same period all over the world, from China and India to Europe and ancient Greece – the ancient Greeks also invented the wheelbarrow. The first picture of a wagon with obvious wheels, dated to 3330 BCE, was dug up in Poland. What seems to have turned potter's wheels into transport and locomotion was the domestication of horses. Some bright spark in one of these places saw the potters at work making drinking vessels, and made the connection with the horses outside. The rest is history, but we will never know who was responsible.

RELATED TOPICS
See also
THE STEAM ENGINE
page 46

THE INTERNAL COMBUSTION ENGINE
page 48

3-SECOND BIOGRAPHY
LEONARD WOOLLEY
1880–1960
English archaeologist who led excavations at Ur in Mesopotamia, where the oldest surviving potter's wheel was found

30-SECOND TEXT
David Boyle

The invention of the wheel as a means of transport marks a turning point in the development of human evolution.

THE NAIL

the 30-second history

Nails were luxury items well into the nineteenth century. In the fledgling United States, unable to access the nails that were made in such quantities in England, people used to make them during the long winter evenings around the fire. The third US president, Thomas Jefferson, used to make nails when he had nothing better to do. Often, when people moved house, they used to remove the nails first because they were so important for holding the joists of their new home together. What made the difference was the invention of 'cut nails', in square shapes made from wrought iron by machine, a process pioneered by the Massachusetts goldsmith Jacob Perkins. Perkins was extraordinarily inventive, pioneering refrigeration and a forerunner of the machine gun, known as the 'steam gun' and rejected by the Duke of Wellington on the grounds that it was 'too destructive'. Perkins had a rival in England called Joseph Dyer, the son of a captain in the Rhode Island Navy, who set up in Birmingham with his own nail-making machine and eventually lost a fortune in the collapse of the Bank of Manchester. Both innovations were overtaken by the development of so-called 'wire nails', which provided us with the cheap, plentiful and diverse nails that we have today.

3-SECOND SURVEY
Nails make it possible to put two pieces of wood together in a way that will hold them in place for a lifetime – or longer.

3-MINUTE OVERVIEW
Who actually invented nails? Nobody knows, but as with so many other pioneering discoveries in this book they seem to have emerged in the Bronze Age, and to have come from the Middle East. The earliest bronze nails found have been carbon-dated back to around 3400 BCE and come from ancient Egypt. They also seem to have been rather expensive.

RELATED TOPICS
See also
IRON
page 22

STEEL
page 24

3-SECOND BIOGRAPHIES
JACOB PERKINS
1766–1849
Pioneering American goldsmith who invented the process of manufacturing cut nails

JOSEPH DYER
1780–1871
American-born inventor, and a manufacturer of cut nails

JOSEPH HENRY NETTLEFOLD
1827–81
British industrialist responsible for bringing wire nails onto the market in vast quantities, and manager of a company now known as GKN

30-SECOND TEXT
David Boyle

Over the millennia nails have been made from bronze, brass, iron, copper and aluminium but today the most commercially viable option is steel.

ARCHIMEDES' SCREW

the 30-second history

3-SECOND SURVEY
Archimedes managed to design a simple pump for the ancient world that is still in use today, particularly for sewage.

3-MINUTE OVERVIEW
Did Archimedes actually think up the screw himself, or did he just see one in action in Alexandria and bring it home? There were reports that the water was raised in this way to irrigate the Hanging Gardens of Babylon, and the classicist Stephanie Dalley has found an Assyrian inscription that seems to push the origin of the screw back to 350 years earlier.

The Archimedes screw – one of the oldest machines still in use – is a device designed to lift low-lying water for drainage or irrigation. More accurately described as a screw pump, the invention is made up of a screw-shaped cylindrical shaft, made of metal or wood, inside a hollow pipe. The screw was turned either by hand or by windmills – the invention was widely used in the Netherlands to drain and reclaim land. As the shaft is turned, the rotating screw shape scoops up water and pushes it out the top of the tube. The Greek engineer and polymath Archimedes is reputed to have come up with the invention to save a ship he had designed. The *Syracusia* was an enormous boat, part luxury vessel, part warship, and big enough to hold a garden, gymnasium and temple to the goddess Aphrodite – but its hold was filling with sea water. Archimedes' simple mechanical pump saved the day, though his ship sailed only once, from Syracuse to Alexandria. Archimedes' design proved to be more long lasting: it is still found in uses as diverse as sewage treatment plants and chocolate fountains and it stabilizes the leaning Tower of Pisa in Italy.

RELATED TOPIC
See also
THE WHEEL
page 34

3-SECOND BIOGRAPHY
ARCHIMEDES
287–212 BCE
Pioneering engineer and mathematician, famous throughout the ancient world, born in Syracuse, Sicily, and murdered there by a Roman soldier

30-SECOND TEXT
David Boyle

The Greek engineer Archimedes was killed for staring at a set of circles he had drawn on the ground rather than obeying a Roman soldier's commands, but his simple pump design lives on.

1 June 1907
Born in Coventry in the English Midlands

27 August 1928
Joins 111 Squadron and wins a reputation for his aerobatic displays

27 January 1936
Enters agreement to set up Power Jets Ltd, formed to develop the first jet

27 August 1939
The first jet-powered plane, the Heinkel He 178, makes its first flight in Rostock

10 December 1940
US patent is applied for; the strain becomes too much for Whittle, who suffers a breakdown and is absent from work for a month

15 May 1941
The first British jet flies, at RAF Cranwell in Lincolnshire; the first flight of the American XP-59A Airacomet follows in October 1942

26 August 1948
Retires from the RAF on medical grounds. He later lives in the United States

9 August 1996
Dies in Columbia, Maryland

1 June 2007
A statue of Whittle is unveiled in his birth town of Coventry

FRANK WHITTLE

The British are famously

innovative but sometimes find it hard to break through an innate conservatism – an adherence to the accepted way of doing things. This means that the story of British invention is often one of brilliant ideas getting stuck. And this was certainly the story of the invention of the jet engine by aircraft engineer Frank Whittle.

Born in 1907 in Coventry, Whittle was the son of a mechanic. He was determined to join the Royal Air Force, but because he measured only 1.5 metres (5 ft) tall he kept on being turned down despite passing the entrance exams with flying colours. He was eventually accepted on his third attempt in 1923.

From his time as an air cadet onwards, Whittle was fascinated by the problems of speed and how to make existing engines go much faster without a necessary increase in the weight and complexity of the propeller engines that were used then. Planes would also need to fly higher to reduce the air pressure, and the existing engine design wasn't suitable for that, either. He came to the conclusion that a gas turbine might provide the necessary power for a major breakthrough in speed.

The result was his design for a turbojet, a fan that sucked in air, squeezed it and set it alight, then pushed it outwards with great power. After the air ministry said it was 'impractical' he patented the idea himself and, because the RAF didn't want it, it was never a state secret. By 1935, he had managed to attract financial backing and started a company called Power Jets Ltd. This was slow going and the official approvals or finance he needed was even slower, while in Germany the Nazis were also pressing ahead with a similar idea. In fact, Whittle was told later that if the German leader Adolf Hitler had known there was a man in England designing planes that could travel at 800 km/h (500 mph), the Second World War might never have happened.

When Whittle's first turbojet flew in May 1941, immediately after the end of the London Blitz, he had been beaten by 21 months by the German pioneer Ernst Heinkel. Five months later, the Bell Aircraft Corporation in America started development of the Airacomet; it was airborne in October 1942, whereas the British Meteor jet did not take to the skies until 1943.

David Boyle

THE PLOUGH

the 30-second history

Probably the man who did most to revolutionize what is, after all, one of the most ancient human inventions of all was a young technician from Berwickshire in southern Scotland called James Small. He was working in Yorkshire when he was 18 and caught sight of one of the iron 'Rotherham ploughs' invented by Joseph Foljambe, who worked in the area and had pioneered the first commercial plough made of metal in 1730. Even so, it took about three decades for his invention to percolate out of the Rotherham region of Yorkshire, where Small saw it. Small took the idea home to Scotland and, in a series of experiments to find the best shape and most appropriate metal, he created what was known as the 'Scots Plough', which was soon exported to the United States. He also refused to patent the idea because he thought his invention should be freely available, though this led at one stage to him spending time in a debtors' prison. Between them, the efforts of Foljambe and Small meant that the business of ploughing a field could be done by a single ploughman and his horse, rather than several men and a team of oxen.

RELATED TOPICS
See also
IRON
page 24

THE STEAM ENGINE
page 48

3-SECOND BIOGRAPHIES
JETHRO TULL
1674–1741
English farmer who invented the seed drill, allowing seeds to be evenly spaced, and launched the 'new husbandry' on the world – much to the horror of his labourers

JOSEPH FOLJAMBE
fl. 1730
English inventor of the first iron plough, or Rotherham plough, the forerunner of the other improvements

JAMES SMALL
1740–93
Scottish perfecter of the metal plough, who refused to patent his discovery

30-SECOND TEXT
David Boyle

Early ploughs relied on human power before the use of draught animals, then machines, to till the soil for crops.

THE MECHANIZED CLOCK

the 30-second history

3-SECOND SURVEY

The mechanized clock revolutionized time-keeping and enabled societies to communicate and cooperate with far greater precision.

3-MINUTE OVERVIEW

The mechanical clock, attached to tolling bells, was not an instrument of liberation in Europe. It was used to impose timetables on ordinary people – when they must get up, when they could break from work in the fields, when they had to go to bed. Time-keeping – and being punctual – became a religious duty, the beginning of a long alliance between puritans and clocks.

Yi Xing, one of the most brilliant inventors in history, was a mathematician, an engineer and a Buddhist monk during the Chinese Tang dynasty. He was also an expert in astronomy and the calendar: he built a series of 13 observatory posts, stretching from Vietnam to Siberia, to perfect his calculations for solar eclipses. Working with a military engineer and government official, Liang Lingzan, his greatest invention was in 725 CE: a mechanical clock that involved water steadily dripping on a wheel that made a full revolution every 24 hours. It combined water power with cogs and wheels and was built in the shape of a great metal sphere that tracked the movements of the stars and planets. The invention may have been a clock, but it did not tell the immediate time. Nor did the eleventh-century mercury-powered astrolabe, invented by a mechanical engineer, al-Murādī, although the device made its way to the Iberian peninsula, where al-Murādī lived, ready to inspire the revolution in mechanical clock-making from around the 1260s onwards. This was when every abbey and cathedral vied with each other to make the most spectacular, ingenious clock mechanisms – mainly powered by more reliable weights. At this time, though, they did not have clock faces: the hours were chimed using bells.

RELATED TOPIC

See also
THE COMPASS
page 54

3-SECOND BIOGRAPHIES

YI XING
683–727
Chinese astronomer who made what seems to have been the first mechanical clock, though still powered by water

LIANG LINGZAN
fl. 720s
Chinese official who worked with Yi Xing to make the first mechanical clock

ALĪ IBN KHALAF AL-MURĀDĪ
fl. c. 1050
Mathematician from Arab Spain who made the first European mechanical clock using mercury

30-SECOND TEXT
David Boyle

Compared with the sundial or hourglass as a means of telling the time, mechanical clocks enabled more accurate time-keeping in the medieval world.

THE STEAM ENGINE

the 30-second history

James Watt was not actually the inventor of the steam engine. Like so many other engineers who emerged from the Scottish Enlightenment, he took an existing idea and made it work effectively. The original inventors were two Englishmen: a military engineer, Thomas Savery, who first invented an 'engine for raising water by fire', and Thomas Newcomen, an ironmonger, whose machine was used for pumping water out of flooded mines. Watt was born in Greenock on the river Clyde in western Scotland. His grandfather was a professor of maths and his father made nautical instruments for the herring fleet. He was hardly poor, but the young Watt still had his difficulties. He was famously moody and difficult to get on with. He began mending and then improving the scale model of Newcomen's engine that Glasgow University used for lectures. He found it was not producing enough steam for it to work efficiently. The piston turned only two or three strokes at a time and most of the steam escaped, and this irritated him. Determined to find ways to prevent the steam in the Newcomen engine being wasted, he made a crucial breakthrough in April 1765, when he realized that steam would rush into a vacuum and need not cool the cylinder, as was the case with Newcomen's engine.

3-SECOND SURVEY
The sheer power of steam made so much else possible. It could drive locomotives, factories, ships, agricultural machinery and much more besides – and soon it did.

3-MINUTE OVERVIEW
The main difference between Watt's and Newcomen's engine was that the steam actually drove the piston rather than creating a vacuum. A few days after Watt's breakthrough, his friend the physicist John Robison barged into his workshop and realized that Watt had solved the problem. But he saw his nervous friend kick the experiment under the table with his foot, terrified that Robison would let the secret out.

RELATED TOPIC
See also
THE INTERNAL COMBUSTION ENGINE
page 48

3-SECOND BIOGRAPHIES
DENIS PAPIN
1647–1713
French physicist who invented a forerunner of the steam engine – the pressure cooker

THOMAS SAVERY
1650–1715
English engineer who built the world's first steam engine

THOMAS NEWCOMEN
1663–1729
English ironmonger who built a steam engine using a vacuum created by condensed steam

JAMES WATT
1736–1819
Scottish inventor whose improvements to the steam engine proved a breakthrough

30-SECOND TEXT
David Boyle

The inventors of the steam engine needed to understand atmospheric pressure and the nature of a vacuum.

Inch.
Vacuum Gauge
No 286 12 17
T A N G Y E
BIRMINGHAM.
Schäffer & Budenberg Ltd
TRADE MARK

THE INTERNAL COMBUSTION ENGINE

the 30-second history

France was the home of the early pioneers of the internal combustion engine – the basic workings of which became the motor car. The invention drives an engine by managing small explosions of gas or petrol in a combustion chamber inside the mechanism and it was developed almost simultaneously by two Frenchmen, one on land and the other on water. The first was a retired soldier, inventor and politician, François Isaac de Rivaz, who was at that time living in Switzerland. It was de Rivaz who first built an engine along these lines, powered by a mixture of oxygen and hydrogen and ignited by an electric spark; he patented the idea in 1807. The following year he built a 'horseless carriage' for the engine to drive. At the same time, Joseph Nicéphore Niépce and his brother Claude were building a similar engine, called the Pyréolophore, fuelled by a mixture of crushed coal dust, moss and resin. This powered a boat on the river Saône and won them a patent the same year (1807). The first modern car was the 1860 brainchild of the Belgian Étienne Lenoir, living at the time in Paris – an adopted Frenchman at least – after which the story passes to Germany.

RELATED TOPIC
See also
THE STEAM ENGINE
page 46

3-SECOND SURVEY
The horseless carriage developed into a must-have element of modern life, which seemed to set people free to come and go as they pleased.

3-MINUTE OVERVIEW
This invention turned out to be a double-edged sword. There are now well over a billion cars on highways of the world, leading to congestion and pollution that shortens the lives of the poorest people, who tend to live closest to the traffic. Nor can the internal combustion engine set people free any more – all too often, it condemns the owners to long, exhausting journeys on congested freeways, breathing in the fumes.

3-SECOND BIOGRAPHIES
FRANÇOIS ISAAC DE RIVAZ
1752–1828
Retired French army officer behind the design for the first hydrogen-powered internal combustion engine

JOSEPH NICÉPHORE NIÉPCE
1765–1833
Pioneer French photographer who developed the first internal combustion engine on a boat, patented in 1807

ÉTIENNE LENOIR
1822–1900
Belgian engineer who built the first commercial motor car engine

30-SECOND TEXT
David Boyle

Efforts to build more environmentally friendly vehicles may spell the end of the internal combustion engine for future transportation of goods and people.

TRANSPORT & LOCATION

THE SAFETY BICYCLE

the 30-second history

3-SECOND SURVEY
The safety bicycle transformed people's ability to travel, particularly working people and women, and became the most important form of personal transport.

3-MINUTE OVERVIEW
The bicycle 'craze' of the 1890s sparked by the Rover's invention had an impact on female emancipation. As bicycles became safer and cheaper, more women had access to the personal freedom the machine embodied. The bicycle came to symbolize the New Woman of the late nineteenth century, especially in Great Britain and the United States. Bicycles were known as 'freedom machines' by feminists and suffragists. They gave women unprecedented mobility and liberated them from restrictive garments.

Early versions of the bicycle in the first half of the 1800s were uncomfortable wooden contraptions with steel wheels and fixed gears. Manufacturers then enlarged the front wheel to ensure a smoother, faster ride, but these 'penny-farthings' were mainly the preserve of thrill-seeking young men who used them for racing. The English inventor John Kemp Starley developed the Rover 'safety' bicycle in 1885. With its lower centre of gravity, this model greatly reduced the danger of 'taking a header' over the handlebars, making braking more effective and increasing the popularity of cycling. Starley's *Rover* was the forerunner of the modern bicycle; its features – near equal-sized wheels, a diamond-shaped frame, pedals below the saddle that power the back wheel through a chain and gears, and handlebars to a forked front wheel – are recognizable today. John Dunlop's reinvention of the pneumatic bicycle tyre in 1888 contributed to a smoother ride on paved streets. At first, cycling was a relatively expensive hobby, but mass production made the bicycle a practical investment for working men getting to and from a job. Women, too, began cycling in increasing numbers, with dramatic changes in their lives. The then-shocking bloomers replaced bustles and corsets, giving a woman more mobility to ride a bike.

RELATED TOPICS
See also
THE WHEEL
page 34

THE PNEUMATIC TYRE
page 58

3-SECOND BIOGRAPHIES
JOHN DUNLOP
1840–1921
Scottish inventor and vet who reinvented the pneumatic tyre for bicycles at his home in Belfast, where he had settled, in 1887

JOHN KEMP (J.K.) STARLEY
1854–1901
English inventor of the revolutionary 'safety' bicycle, on which modern-day bicycle technology is still based

30-SECOND TEXT
Judith Hodge

Cycling, first made popular when the safety bicycle replaced earlier models known as 'bone-shakers', is today experiencing a resurgence.

THE PNEUMATIC TYRE

the 30-second history

The pneumatic tyre was first developed by Scottish inventor Robert Thomson. He was only 23 when he patented his invention in 1846 – a hollow belt of India rubber inflated with air so that the wheels presented 'a cushion of air to the ground, rail or track on which they run'. The belt of rubberized canvas was encased within hard-wearing leather that was bolted to the wheel. A demonstration of Thomson's 'Aerial Wheels' fitted to several horse-drawn carriages took place in Regent's Park, London, in March 1847. The pneumatic tyres greatly improved both passenger comfort and noise levels, and one set was reported to last for 1,950 kilometres (1,200 miles). But Thomson's invention was ahead of its time; not only were there no motor cars, but bicycles were only just starting to appear. Lack of demand and high production costs meant that pneumatic tyres remained a mere novelty for another 30 years, until their reinvention by John Boyd Dunlop. Thomson developed and patented many other inventions, including the first successful mechanical road haulage vehicle, a steam traction engine. His solid India-rubber tyres (another Thomson patent) enabled heavy steam engines to travel along roads without damaging the surface. By 1870, 'Thomson Steamers' were being manufactured and exported around the world.

RELATED TOPICS
See also
THE SAFETY BICYCLE
page 56

THE AUTOMOBILE
page 62

3-SECOND BIOGRAPHY
ROBERT WILLIAM THOMSON
1822–73
Scottish inventor who patented the pneumatic tyre in 1846

30-SECOND TEXT
Judith Hodge

3-SECOND SURVEY
Invention of the pneumatic tyre, providing a cushion of air between road and vehicle, played a key role in the boom in bicycles and automobiles.

3-MINUTE OVERVIEW
The name of another Scottish-born inventor, John Boyd Dunlop (1840–1921), is more closely associated with the invention and the company that still bears his name. Familiar with making rubber devices, he reinvented pneumatic tyres for his son's tricycle and developed them for use in cycle racing. After a cyclist using his tyres started winning all the races, Dunlop sold his rights to Dunlop Rubber, a company formed by president of the Irish Cyclists' Association, Harvey Du Cros.

Overlooked as the inventor of pneumatic tyres, a vintage vehicle rally is held in honour of Robert Thomson every June at his Scottish birthplace, Stonehaven.

16 April 1867
Wilbur born in Milville, Indiana

19 August 1871
Orville born in Dayton, Ohio

1889–92
The brothers launch several newspapers

1892
They open bicycle repair and sales shop, selling own bicycle design

1900
They carry out first tests with glider at Kitty Hawk

17 December 1903
Wilbur Wright makes the first controlled, sustained powered flight, in the *Wright Flyer I*

1905
The brothers develop and demonstrate the first practical aeroplane

1906
Patent granted on aeroplane control system

1908
Public demonstrations in Europe and North America. Orville crashes and is badly injured; his passenger becomes the first person to die in an aeroplane crash

1909
US Army buys its first military aircraft. The Wrights begin to manufacture aeroplanes and teach pilots; they become wealthy businessmen

30 May 1912
Wilbur dies in Dayton of typhoid, aged 45

1915
Orville sells the business

30 January 1948
Orville dies in Dayton, aged 76

THE WRIGHT BROTHERS

It was a childhood toy, a model helicopter, brought home by their preacher father that sparked Wilbur and Orville Wright's lifelong interest in mechanics and aviation. The brothers worked together on various enterprises, including newspaper publishing and running a bicycle shop in their hometown, Dayton, Ohio. They were swept up in the cycling craze following the invention of the safety bicycle, even designing their own model.

The Wrights followed the research of early aviators such as Otto Lilienthal, who died in a glider crash. Experimenting with 'wing warping' (a system they later patented of pulleys and cables to control a fixed-wing aircraft) based on bird flight, the brothers built and tested their flying machines at Kitty Hawk, North Carolina, known for its strong winds. It was here on 17 December 1903 that they made the first controlled, sustained flight of a powered, heavier-than-air aircraft. Wilbur flew the *Wright Flyer I* for 59 seconds, at 260 metres (852 feet), recorded in a famous 'first flight' photograph.

The achievement did not bring them overnight fame and fortune. Instead, news of the event was met with great scepticism both at home and abroad: the French press dismissed them as 'bluffeurs'. Over the next two years, the brothers developed the first practical fixed-wing aeroplane, which they demonstrated in 1905. Although not the first to fly experimental aircraft, the Wrights invented the controls that made flight possible, and they patented their design in 1906.

Unable to find a market for their invention, Wilbur went to Europe where, during public demonstrations in France, he stunned onlookers with the plane's abilities and his skills as a pilot, performing figures of eight. Back in the United States, similar displays by Orville resulted in a crash, leaving him badly injured and his passenger, Thomas Selfridge, dead. Shocked by the event, Wilbur pushed himself to greater lengths and set new records for altitude and duration.

The Wrights became huge celebrities, feted by royalty and heads of state, as well as wealthy businessmen with sales of their planes in Europe and the United States. Wilbur died in 1912, and Orville sold the business in 1915. The remaining Wright brother made his last flight in 1944, noting that the plane had a wingspan longer than the distance of their first flight in 1903.

Judith Hodge

THE AUTOMOBILE

the 30-second history

Karl Benz is generally credited

with the invention of the modern automobile powered by an internal combustion engine. The Benz Patent-Motorwagen went into production in 1886, after Benz's wife Bertha made the first successful road trip from Mannheim to Pforzheim and back in the 'horseless coach'. Close behind came fellow German inventors Gottlieb Daimler and Wilhelm Maybach; in 1889 they designed the first vehicle from scratch as an automobile, rather than a horse-drawn carriage fitted with an engine. Before this, earlier self-propelled vehicles included Nicolas-Joseph Cugnot's steam-powered tractor for the French army (1769), and Robert Anderson's electric carriage (c. 1832–9). The focus of British inventors was on rail after the Locomotive Act (from 1865 to 1896, self-propelled vehicles on public roads had to be preceded by someone waving a red flag) effectively killed off auto development. It was in the United States that cars were rapidly adopted in the early twentieth century. Henry Ford's mass-production techniques led to the first affordable car for the general public, the 1908 Model T. By the 1930s, most of the mechanical technology used in cars today had been invented, although some of it was later reinvented. Future automobile inventions are focused on energy efficiency, alternative fuels and auto-pilot systems.

3-SECOND SURVEY
The so-called Automobile Age, particularly in the United States, has revolutionized the way people live, work and play – and fuelled huge social and economic changes.

3-MINUTE OVERVIEW
The first electric car was the 1888 Flocken Elektrowagen by German inventor Andreas Flocken. Electric cars were popular at the turn of the nineteenth century, but the advantages of petroleum cars – greater range, quicker refuelling and finally the Ford company's mass production (petrol cars became half the price) – led to a decline. But current concerns about carbon emissions and dwindling fuel resources have reignited interest in hybrid vehicles and electric cars.

RELATED TOPICS
See also
PETROLEUM
page 26

THE WHEEL
page 34

INTERNAL COMBUSTION
ENGINE
page 48

3-SECOND BIOGRAPHIES
NICOLAS-JOSEPH CUGNOT
1725–1804
French inventor of first self-propelled vehicle, a steam-powered military tractor

KARL FRIEDRICH BENZ
1844–1929
German inventor of first petrol automobile

30-SECOND TEXT
Judith Hodge

Ford is credited with producing the first car that gave the middle class the means to travel independently, the Model T, known affectionately as the Tin Lizzie.

THE HELICOPTER

the 30-second history

The first helicopter was imagined by Renaissance artist and inventor Leonardo da Vinci, whose drawings of a 'helix' flying machine were inspired by bird flight and the floating maple seed. But the modern helicopter had to wait for advances in aerodynamic theory, structural materials and more powerful engines in order to get off the ground. Igor Sikorsky's first helicopter, built in 1909, was made of wood and only had a 25 horsepower engine. Thirty years later, he invented the successful VS-300, which went on to become the model for all modern single-rotor helicopters. The invention of the helicopter stems from the evolution of a simple idea: unrestricted flight – moving vertically and horizontally, or hovering in mid-air. Unlike a fixed-wing plane, the helicopter is capable of vertical lift off and landing – thanks to its rotary blade. It can also hover in a fixed position. These attributes make it ideal wherever space is limited or for anything where it has to stay over a precise area. That means we can use helicopters in anything from rescue work to firefighting, aerial photography and filming, accessing remote locations for environmental or relief work, and to deliver supplies and workers to remote oil rigs – all those uses, despite the fact that helicopters were originally invented solely for military and intelligence purposes.

RELATED TOPIC
See also
THE WRIGHT BROTHERS
page 60

3-SECOND BIOGRAPHIES
LEONARDO DA VINCI
1452–1519
Italian 'Renaissance man', a gifted artist, mathematician, scientist and designer-inventor

IGOR SIKORSKY
1889–1972
Russian-American designer of aeroplanes, flying boats and the first viable helicopter

30-SECOND TEXT
Judith Hodge

3-SECOND SURVEY
The invention of the helicopter transformed flight. Its unique design meant it was capable of vertical lift off and had the ability to hover over a single spot.

3-MINUTE OVERVIEW
The helicopter was put to military use almost immediately. Sikorsky's R-4 model was the only helicopter to serve in the Second World War, where it was used as a direct lift aircraft to rescue people trapped in areas inaccessible by planes. But it was the difficult terrain of the Korean and Vietnam wars where military helicopter use became widespread. Further technological refinements have made it a mainstay of modern warfare.

In spite of the contributions of other engineers to the conception and development of the helicopter, Sikorsky's name is synonymous with the invention.

GLOBAL POSITIONING SYSTEM (GPS)

the 30-second history

The global positioning system (GPS) was originally designed for military and intelligence use. The US government developed GPS at the height of the Cold War in the 1960s to track US submarines carrying nuclear missiles. Scientists found that satellites could be tracked from the ground by measuring the frequency of the radio signals they emitted, the so-called 'Doppler Effect'. GPS uses a network of satellites that orbit Earth and beam down signals carrying a time code and geographical data to a GPS receiver. The invention has transformed navigation, since GPS accurately calculates geographical positions to a matter of metres in all weather conditions, and operates independently of telephone or Internet reception. But it wasn't until 1983, when the former Soviet Union shot down a Korean passenger jet that had strayed into restricted airspace, that the US government opened up GPS for civilian use so that aeroplanes, shipping and transport could accurately fix their positions. The US government maintains the system and makes it freely accessible, but it can selectively deny GPS access, for example to the Indian military in 1999 during hostilities in Kashmir. Today, GPS has many uses beyond navigation such as map-making, earthquake research, climate studies and an outdoor treasure-hunting game known as geocaching.

RELATED TOPIC
See also
THE COMPASS
page 54

3-SECOND BIOGRAPHY
ROGER LEE EASTON
1921–2014
American scientist and physicist, principal inventor and designer of the Global Positioning System, along with Ivan Getting and Bradford Parkinson

30-SECOND TEXT
Judith Hodge

3-SECOND SURVEY
GPS is the most significant development in navigation since the introduction of radio navigation during the late 1940s.

3-MINUTE OVERVIEW
Due to the potential denial of access and monitoring by the US government, other alternative systems are in use or being developed. These include the Russian Global Navigation Satellite System (GLONASS), which can be added to GPS devices, making more satellites available and enabling positions to be fixed to within 2 metres (6½ feet). Others include the planned European Union Galileo positioning system, China's BeiDou Navigation Satellite System and India's Indian Regional Navigation Satellite System (NAVIC).

GPS receivers have been miniaturized to a few integrated circuits; they are now found in cars, planes and boats, as well as laptop computers.

MEDICINE & HEALTH

adrenaline Hormone produced by the body that increases rates of blood circulation and breathing and makes the heart beat faster, preparing it to react to danger (the 'fight or flight response').

anthrax Serious, often fatal bacterial disease of sheep, cattle and other mammals. It can be transmitted to people via contaminated wool, raw meat or other animal products, typically affecting the skin and lungs.

antimicrobial resistance Resistance that occurs when microorganisms such as bacteria and viruses change in ways that render the medications used against them ineffective.

bacteria Single-celled microscopic organisms found everywhere in Earth's environment; some species are harmless or even beneficial, such as those involved in fermentation or decomposition, while others produce disease-causing toxins.

bite the bullet Expression that originates from a time before effective anaesthetics, when soldiers were given bullets to bite on to help them endure pain; figurative usage means to endure a painful or difficult situation seen as unavoidable.

Boer War Conflicts between the British and the South African Boers; the first (1880–1) when the Boers revolted against the 1877 British annexation of the Transvaal, and the second (1899–1902) fought between the British and the two independent Boer republics of the Orange Free State and Transvaal.

chickenpox Highly contagious disease most common in children, caused by herpes virus and characterized by itchy, inflamed blisters that eventually scab over.

chloroform Colourless, volatile liquid with a sweet taste and odour, used as a solvent and in refrigerants; formerly used as an anaesthetic.

cowpox Contagious skin disease of cattle, usually affecting the udder, caused by a virus; resembles mild smallpox when contracted by humans through contact, and was the basis of the first smallpox vaccines.

diabetes Disease in which the body's ability to produce or respond to the hormone insulin is impaired, resulting in elevated levels of glucose in the blood and possible health complications, including heart disease and blindness.

diphtheria, tetanus and pertussis (DTP) immunization Combined vaccine to immunize children. Its side effects led to the development of safer acellular vaccines (DTaP).

electromagnetic radiation Form of energy associated with electric and magnetic fields that includes visible light, radio waves, microwaves, gamma rays and X-rays.

ether Highly flammable, colourless liquid used as a solvent in industrial processes; formerly used as an anaesthetic.

hepatitis Inflammation of the liver, most commonly caused by hepatitis virus; there are several types, including hepatitis A, typically caused by contaminated food or water, and hepatitis B, transmitted by infected body fluids such as blood.

HIV Human Immunodeficiency Virus, which damages the immune system; if untreated, HIV can result in the potentially life-threatening condition AIDS (Acquired Immunodeficiency Syndrome).

malaria Serious infectious disease of humans (and other animals) spread by certain mosquitoes, characterized by recurrent symptoms of chills, fever and an enlarged spleen; common in tropical climates and endemic in many developing countries. There is currently no vaccine available.

measles, mumps and rubella (MMR) vaccine Combined vaccine to immunize children; safety concerns involving the vaccine, including a recent controversy over possible links to autism, have been disproved.

meningitis Disease involving inflammation of the membranes (meninges) surrounding the brain and spinal cord, caused by viral or bacterial infection.

nitrous oxide Colourless, odourless gas used in surgery and dentistry as a mild anaesthetic; known as 'laughing gas' due to the euphoric effects of inhaling it, a property that has led to its recreational use.

rabies Contagious and potentially fatal viral disease of dogs and other mammals, transmissible through the saliva to humans usually by the bite of infected animals, and causing madness and convulsions.

streptococcus Group of bacteria that causes illnesses such as strep throat, strep pneumonia, scarlet fever and rheumatic fever.

thimerosa Mercury-containing preservative to prevent growth of bacteria or fungi; no longer used in vaccines for babies and young children due to safety concerns.

tuberculosis (TB) Infectious disease caused by bacteria, characterized by the growth of nodules mainly in the lungs; it can be treated, but drug-resistant TB is increasing.

VACCINATION

the 30-second history

Vaccines are often described as one of the greatest achievements of public health. It is estimated that widespread vaccination has reduced morbidity by 99 per cent for five 'killer' diseases: smallpox, diphtheria, measles, polio and rubella. Smallpox was a particular scourge for centuries; a basic form of vaccination, known as variolation, was practised in China and India as early as 900 CE: healthy people were exposed to powdered scabs from smallpox pustules inserted either under the skin or in the nostrils. An English doctor, Edward Jenner, is generally recognized as the inventor of the first 'modern' vaccine. Jenner observed that people who had suffered from the less virulent cowpox seemed immune to smallpox. In 1796, he carried out his famous experiment on eight-year-old James Phipps, inserting pus from a cowpox pustule into the boy's arm. Neither James or the other children he vaccinated developed smallpox (Jenner took the name vaccine from the Latin *vacca* for cow). Despite opposition from both the medical profession and the Church, the invention was widely adopted in Europe and the United States, and further developed to protect against other diseases. Louis Pasteur developed vaccines for anthrax and rabies in the nineteenth century, using 'killed' agents rather than live viruses.

RELATED TOPICS

See also
ANTIBIOTICS
page 76

THE MICROSCOPE
page 78

LOUIS PASTEUR
page 80

3-SECOND SURVEY

Vaccination may have eradicated diseases such as smallpox, although vaccines for other 'plagues' such as malaria and HIV prove more elusive.

3-MINUTE OVERVIEW

The 1956 WHO vaccination campaign to eradicate smallpox was declared effective in 1980, after the last naturally occurring case of smallpox was reported in Somalia in 1977. Viruses such as polio have remained more resistant to vaccination efforts, although the disease is now rare. Controversy has surrounded the safety and efficacy of the diphtheria, tetanus and pertussis (DTP) immunization and the measles, mumps and rubella (MMR) vaccine; as well as the use of a mercury-containing preservative called thimerosal.

3-SECOND BIOGRAPHIES

EDWARD JENNER
1749–1823
English doctor who invented first 'modern' vaccine for smallpox

LOUIS PASTEUR
1822–95
French microbiologist, famous for his discoveries of the principles of vaccination and pasteurization

MAURICE HILLEMAN
1919–2005
American microbiologist and most prolific vaccine inventor

30-SECOND TEXT
Judith Hodge

Vaccinations have been widely available since the 1920s; many of those currently licensed were invented by Hilleman.

ANAESTHESIA

the 30-second history

The search for an effective method to control pain, particularly for surgical operations, took centuries to resolve. Alcoholic concoctions were used as early anaesthetics, as were plants containing powerful chemicals. A breakthrough came in 1846, when American dentist William Morton instructed a patient to inhale the fumes of an ether-soaked sponge in front of doctors and students at Massachusetts General Hospital. Once the patient was sedated, the surgeon, John Warren, removed a tumour from the patient's neck. When he awoke, the patient reportedly declared that 'it feels as though my neck's been scratched'. Up until then, surgery was quick and brutal, with fully conscious patients having to bite the bullet – sometimes quite literally – during the operation. Although ether had been known of for about 300 years, no one had thought to use it as an anaesthetic. It had instead been used as a recreational drug at parties known as 'ether frolics'. News of Morton's success quickly spread and numerous operations using ether were performed by respected European surgeons. Anaesthetic obviously made surgery a better experience for patients, but it also allowed surgeons more time when operating, which in turn has led to advances in surgical techniques. In time, ether was replaced by safer and more reliable drugs.

3-SECOND SURVEY
Anaesthesia gives temporary relief from pain so that surgeons can do their work quickly – without their patients dying from shock.

3-MINUTE OVERVIEW
Modern chemistry led to the development of isolated anaesthetics. Of these early drugs, only nitrous oxide (laughing gas) is still widely used today. Both ether (which is highly flammable and causes excessive vomiting) and chloroform (which is pharmacologically dangerous) have been replaced by safer general anaesthetics. Current anaesthesia practices involve the use of one or more drugs tailored to the particular patient and procedure, although there are still major risks such as heart attack or pulmonary embolism.

RELATED TOPICS
See also
VACCINATION
page 72

ANTIBIOTICS
page 76

3-SECOND BIOGRAPHIES
WILLIAM MORTON
1819–68
American dentist who first demonstrated the use of ether as a surgical anaesthetic in 1846

JOHN SNOW
1813–58
English physician who popularized obstetric anaesthesia by chloroforming Queen Victoria for the birth of Prince Leopold, her eighth child, in 1853

30-SECOND TEXT
Judith Hodge

The field of anaesthetics has developed into a highly skilled medical practice since the experiments made by early practitioners such as Morton and Snow.

ANTIBIOTICS

the 30-second history

Antibiotics revolutionized medicine in the twentieth century. Before 1935, people routinely died from minor cuts and scratches that had become infected. Penicillin, discovered in 1928, is often cited as the first antibiotic but drugs containing sulphur were being used to treat bacterial infections a decade earlier. The bright orange-red compound prontosil was used as an industrial dye before German scientist Gerhard Domagk started testing it on mice injected with a lethal dose of *streptococcus*. The mice given prontosil survived. The biggest test for Domagk came when his young daughter developed a *streptococcus* infection in 1935; despite minimal testing on humans, he made the desperate decision to give her large oral doses of prontosil and, two days later, she was miraculously cured. Domagk was given the Nobel Prize for his discovery in 1939. But penicillin is still the world's most famous antibiotic. Scottish biologist Alexander Fleming noticed that mould had developed on dishes being used to grow the *staphylococci* germ – and had created a bacteria-free circle around itself. But he needed to do more work to develop penicillin into a mass-produced drug. Penicillin was in big demand during the Second World War, when it was widely used to treat infected wounds and diseases such as syphilis.

3-SECOND SURVEY
The advent of antibiotics hailed a new era in the treatment of communicable disease because it meant that people no longer died from common infections.

3-MINUTE OVERVIEW
The effectiveness and accessibility of antibiotics has led to their overuse, particularly as growth promoters in animals and fish. Microbial resistance results in 'superbugs' that no longer respond to antibiotics, now seen in forms of tuberculosis and increasingly in HIV and malaria. In 2014 the World Health Organization (WHO) classified antimicrobial resistance as a 'serious threat' affecting every region of the world. Researchers are currently exploring alternatives to antibiotics, including predatory bacteria and metals.

3-SECOND BIOGRAPHIES
ALEXANDER FLEMING
1881–1955
Scottish biologist whose accidental discovery of penicillin is widely heralded as the start of modern antibiotics

GERHARD DOMAGK
1895–1964
German bacteriologist credited with creating the first commercial antibiotic

30-SECOND TEXT
Judith Hodge

Hailed as one of the "wonder drugs," of the last century, few new families of antibiotics have been created since the early 1960s. Fleming himself cautioned against penicillin misuse.

THE MICROSCOPE

the 30-second history

The Dutch scientist Anton van Leeuwenhoek is credited with being the first person to make and use a practical microscope. Before his late seventeenth-century innovations, the ancient Romans and Greeks and the Chinese had used lenses for magnification. In the 1590s, two Dutch spectacle-makers, the Janssens, put several lenses in a tube and discovered that an image magnified by a single lens can be further magnified by a second or further lenses. With a 'compound' microscope like this English scientist Robert Hooke made observations and drawings of the hairs on the bodies of fleas, captured in his 1665 book *Micrographia*. He also discovered pores in thin slices of cork, calling them plant 'cells' because they reminded him of cells in a monastery. Leeuwenhoek achieved greater success with magnification by developing new ways of grinding lenses to improve the optical quality – up to 200x magnifying power when other microscopes of the time had 20–30x magnification. His hand-held microscope used a single convex glass lens attached to a metal holder and was focused using screws. In a famous experiment in 1674, he viewed the teeming, previously unseen life in a drop of lake water. He was the first person to describe bacteria, yeast plants and the circulation of blood corpuscles in capillaries.

3-SECOND SURVEY
Before the microscope, people did not realize that living things were made up of tiny structures that could not be seen with the naked eye.

3-MINUTE OVERVIEW
Optical microscopes are only able to focus on objects that are at least the size of a wavelength of light. The invention of the electron microscope in 1931 by Max Knoll and Ernst Ruska shattered the limitations of visible light by transmitting a beam of electrons (instead of light) through a specimen. Ruska's principles still form the basis of modern electron microscopes – microscopes that can achieve magnification levels of up to two million times.

RELATED TOPICS
See also
THE OPTICAL LENS
page 82

THE X-RAY
page 84

3-SECOND BIOGRAPHIES
ZACHARIAS JANSEN
c. 1580–c. 1638
Dutch spectacle-maker, who, with his father Hans, invented the first microscope

ANTON VAN LEEUWENHOEK
1632–1723
Dutch draper and scientist, 'father' of the microscope

ERNST RUSKA
1906–88
German physicist who won the Nobel prize in 1986 for his work on electron optics including invention of electron microscope

30-SECOND TEXT
Judith Hodge

Great advances in microscopy had been made by the mid-1600s, after which microscopes remained unchanged for the next 200 years.

27 December 1822
Born in Arbois, third child of a poor tanner Jean-Joseph Pasteur, and Jeanne-Etiennette Roqui

1848
Discovers 'isomerism', existence of left- and right-handed molecules as chemist at University of Strasbourg

29 May 1849
Marries Marie Laurent: they have five children, only two of whom survive to adulthood

1859
Develops controversial 'germ theory', through experiments with fermentation

1862
Wins prize for disproving theory of spontaneous generation; invents pasteurization

1865–67
Rescues European silk industry through discovery of microbes attacking silkworm eggs

1868
Suffers stroke, aged 45

1879
Works on chicken cholera, and man-made vaccines

1881
Develops vaccine for anthrax

6 July 1885
Vaccinates a nine-year-old boy, Joseph Meister, against rabies

1887
Founds the Pasteur Institute; remains director until his death

28 September 1895
Dies at 72 from another stroke

LOUIS PASTEUR

As a child, Louis Pasteur showed more interest in painting and drawing than science. His first discovery, as a chemist at the University of Strasbourg, was that almost identical molecules could exist as left- and right-handed 'mirror' versions. This had huge implications for microbiology (of which Pasteur is called the 'father') and drug development.

In 1849, he married Marie Laurent, who assisted him in all his scientific experiments. Only two of their five children survived to adulthood, the others dying of typhoid. Pasteur wrote that such personal tragedies motivated him to find cures for infectious diseases.

A breakthrough came when he developed the controversial 'germ theory'. In 1862 the French Academy of Sciences offered a prize of 2,500 francs to anyone who could prove or disprove that life was spontaneously generated. In his most famous experiment, Pasteur demonstrated that food was contaminated due to microbes in the air but could be sterilized by heating. He achieved fame with his discovery when the theory was put into practice.

France's wine trade was suffering huge losses from bottles spoiled in transit. The invention of pasteurization – in this case, briefly heating wine to 55°C (131°F) – killed off microorganisms without affecting the taste. The process was extended to other farm and brewery products such as milk and beer. Beverage contamination led Pasteur to the idea that microorganisms in animals and humans might also be causing disease, and his research inspired British surgeon Joseph Lister to develop antiseptic methods in surgery.

Despite not being a biologist, Pasteur accepted the challenge to save the European silk industry from an unknown disease. By 1868, he had discovered that not one but two parasitic microbes were attacking the worms' eggs, and that disinfecting the nurseries would solve the problem.

Pasteur suffered a stroke at 45, probably due to overwork, but a mobile laboratory was set up at his bedside. Through his study of chicken cholera he built on Edward Jenner's earlier work, and manufactured vaccines for both rabies and anthrax. However, recent studies of his notebooks reveal that Pasteur may have overstated how much original work he did. He secretly used the method of veterinarian Jean Toussaint to prepare an anthrax vaccine. The first human trial of a rabies vaccine, when Pasteur vaccinated a boy who had been bitten by a rabid dog, was a success but he had exaggerated how much animal testing he had carried out beforehand.

In 1887 Pasteur founded the research institute that bears his name. Although he achieved many things, Pasteur is probably best remembered for the invention that commemorates his name, pasteurization.

Judith Hodge

THE OPTICAL LENS

the 30-second history

The first lenses, dated as early as 700 BCE, were polished rock crystals. The Roman emperor Nero is said to have watched gladiators using an emerald as a corrective lens. Two early Arabic scientists transformed how light and vision were understood. Ibn el-Haitam was the first to correctly describe how light is refracted by a lens (and how the eye functions): previously people had believed that light was released rather than reflected by objects. Ibn al-Haytham (Alhazen) wrote about the effects of concave lenses and magnifying glasses in his *Book of Optics*. The book was translated into Latin in the twelfth century and inspired Franciscan monk Roger Bacon to recommend glass spheres as magnifying glasses to help people read. By the thirteenth century, Venice and Florence were home to an optical industry, grinding and polishing lenses. It was in Italy that the first wearable eyeglasses (using lenses for both eyes) were invented at the end of the 1280s. Literacy was not widespread in the fourteenth century, but the invention of printing led to a huge increase in books and pamphlets – and a corresponding rise in reading and need for reading glasses. Further development of the optical lens by Dutch spectacle makers led to the invention of the microscope and the refracting telescope in 1608.

RELATED TOPICS
See also
THE MICROSCOPE
page 78

THE PRINTING PRESS
page 98

3-SECOND BIOGRAPHIES
IBN AL-HAYTHAM (ALHAZEN)
c. 965–c. 1040
Arabic scientist, author of influential *Book of Optics*

ROGER BACON
c. 1214–94
English Franciscan monk and philosopher whose study of the optical properties of lenses and mirrors led to charges of 'witchcraft' against him

GALILEO GALILEI
1564–1642
Italian polymath who developed the Galilean telescope and mapped the night sky

30-SECOND TEXT
Judith Hodge

Eyeglasses with optical lenses were first worn in 1284 in Italy, replacing 'reading stones' made by cutting a glass sphere in half.

THE X-RAY

the 30-second history

The invention of the X-ray in 1895 meant that for the first time the inner workings of the body could be made visible without surgery. While studying the effects of passing an electrical current through gases at low pressure, German physicist Wilhelm Röntgen accidentally discovered that electromagnetic radiation was capable of penetrating most solid objects. He named these X-rays ('X' stood for unknown), and later sent an X-ray photograph of his wife's hand – with bones and a wedding ring clearly visible – to colleagues. The discovery transformed medicine almost overnight. Within a year, the first radiology department had opened at Glasgow Hospital, producing the first pictures of a kidney stone and a penny lodged in a child's throat. Shortly after, X-rays were used to trace the path of food through the digestive system. Röntgen's discovery was used in the Boer War and in the First World War, to locate bone fractures and embedded bullets in wounded soldiers. There was much public excitement at the new technology, and X-ray machines appeared as a wondrous curiosity in theatrical shows. Today, X-rays are most widely used in medicine to analyse problems with bones, teeth and organs in the body. They also have uses in industry for tasks such as detecting cracks in metal, and security checks at airports.

RELATED TOPICS

See also

ANAESTHESIA
page 74

ANTIBIOTICS
page 76

THE MICROSCOPE
page 78

3-SECOND SURVEY
Röntgen's accidental discovery of the X-ray rapidly transformed medicine by enabling people to see the inside of the human body without cutting into flesh.

3-MINUTE OVERVIEW
The widespread and unrestrained use of X-rays led to serious injuries, with people who were X-rayed or who worked with the original machines suffering radiation burns, hair loss and organ damage. Early dosages were up to 1,500 times higher than those used today. It was eventually recognized that frequent exposure to X-rays could be harmful; today special measures are taken to protect the patient and doctor. High doses of X-rays are used to treat cancer.

3-SECOND BIOGRAPHY
WILHELM RÖNTGEN
1845–1923
German physicist awarded the first Nobel Prize in physics in 1901 for his discovery of X-rays, an invention that continues to be used in hospitals worldwide

30-SECOND TEXT
Judith Hodge

X-rays are used in other branches of science such as crystallography, which was used by Rosalind Franklin to discover the structure of DNA, and microscopic analysis, which uses X-rays to produce images of very small objects.

THE BIRTH CONTROL PILL

the 30-second history

Inextricably linked to free love, the 'swinging sixties' and women's liberation, the contraceptive pill was developed in the 1950s by American biologists Gregory Pincus and John Rock. Arguably, few inventions have had such a profound social impact in terms of transforming women's lives – though the pill's road to widespread use has been a rocky one. At a New York dinner party, birth control activist Margaret Sanger persuaded Pincus to work on a contraceptive pill, a combination of the hormones oestrogen and progesterone. In 1956 large-scale clinical trials conducted in Puerto Rico deemed the pill to be 100 per cent effective, but some serious side effects were ignored. The following year the US Federal Drug Administration (FDA) approved the pill, but only for severe menstrual disorders – an unusually large number of women reported severe menstrual disorders. When it was approved for contraceptive use in 1960, take-up was rapid: within five years, it was the most popular form of birth control in the United States, used by 6.5 million American women. It was introduced in the UK on the NHS in 1961 for married women only – this lasted until 1967 – and is now taken by 3.5 million women in Britain between the ages of 16 and 49. The pill is currently taken by about 100 million women worldwide.

3-SECOND SURVEY
The pill was one of the most significant medical advances of the twentieth century, since it changed the way that sexual relationships were viewed.

3-MINUTE OVERVIEW
Health scares have dogged the pill from the beginning, with reports linking its use with blood clots, strokes, heart attacks and diabetes. Research in the 1980s suggested links between pill use and breast cancer, causing user numbers to dip. Some of the concerns were linked to hormone levels in the pill, which have now been lowered. On the plus side, the pill has been shown to be protective for ovarian cancer and pelvic inflammatory disease.

RELATED TOPICS
See also
VACCINATION
page 72

ANTIBIOTICS
page 76

3-SECOND BIOGRAPHIES
MARGARET SANGER
1879–1966
American birth control activist who opened the first birth control clinic and started the Planned Parenthood Federation of America, which funded the pill's development

JOHN ROCK & GREGORY PINCUS
1890–1984 & 1903–67
American co-inventors of the pill

CARL DJERASSI
1923–2015
Bulgarian-American chemist who chemically created the pill by synthesizing Mexican yams in 1951 but was unable to manufacture or distribute it

30-SECOND TEXT
Judith Hodge

The pill was the first-ever lifestyle drug; it gave women control over their own fertility.

Start

THE PACEMAKER

the 30-second history

The invention of the pacemaker, a device that emits electrical pulses that prompt the heart to beat at a normal rate, was literally life-saving. Before its advent there were only two methods to help someone with irregular heart rhythms or suffering cardiac arrest: mechanical stimulation of the heart or injecting a drug such as adrenaline through the chest wall. Both were unreliable. Early pacemakers were external machines as big as TV sets and gave electric shocks when used. Albert Hyman's 1932 invention was powered by a hand-cranked motor. Public perception that pacemakers were interfering with nature by 'reviving the dead' may have curtailed research until the 1950s. In 1958, Arne Larsson became the first person to receive an implantable pacemaker attached by electrodes to the heart's muscle tissue. The device was designed by Swedish doctor and inventor Rune Elmqvist, working closely with cardiac surgeon Åke Senning. The patient went on to receive 26 different pacemakers during his lifetime and died at 86, having outlived both inventor and surgeon. The next important innovation was overcoming the short lifetime of early pacemaker batteries. In 1971, Wilson Greatbatch invented the long-life lithium-iodine battery, which became the standard for future pacemaker designs.

3-SECOND SURVEY
The invention of the pacemaker has saved thousands of lives, using electrical pulses to regulate the beating of the heart.

3-MINUTE OVERVIEW
Modern pacemakers have external programming tailored for individual patients so that a cardiologist can select the optimum pacing modes. Some combine a pacemaker and defibrillator in a single implantable device. Others have multiple electrodes stimulating differing positions within the heart to improve the way the lower chambers (ventricles) of the heart are synchronized. The latest pacemaker technology is a new wireless device about the size of a grain of rice, inserted directly into the heart.

RELATED TOPIC
See also
THE X-RAY
page 84

3-SECOND BIOGRAPHIES
ALBERT HYMAN
1893–1972
American cardiologist who, along with brother Charles, built one of the earliest 'artificial pacemakers' – he was the first to use the term

RUNE ELMQVIST
1906–96
Swedish doctor, engineer and inventor who designed the first implantable pacemaker

WILSON GREATBATCH
1919–2011
American engineer and inventor; holder of 350 patents, including the lithium-iodine battery used as standard cell for pacemakers

30-SECOND TEXT
Judith Hodge

By the mid-1980s, pacemakers – now the size of a large coin – were routinely fitted to people with heart conditions.

THE ALPHABET

the 30-second history

The ancient Greeks created the first true alphabet in the eighth century BCE. Their breakthrough innovation was to use characters to represent vowels, completing a process that had begun thousands of years earlier, and producing a writing system that could record speech without ambiguity and was accessible to ordinary people. Archaeological evidence suggests that by around 1700 BCE Semitic workers had already simplified the alphabet from the unwieldy hieroglyphs of the ancient Egyptians. Reducing the number of characters to a couple of dozen symbols, 'Proto-Sinaitic' script was useful for commerce, which led to it becoming widespread, with many variants created. The Phoenician thalassocracy of the eighth century BCE created the phonemic alphabet by adapting characters to represent consonants rather than syllables. Trading throughout Asia Minor and the Mediterranean, the Phoenicians used their consonantal alphabet to record any language they encountered. Since the transactions they recorded related to money, there was huge motivation among their trading partners to learn Phoenician script, thereby spreading its use. But the lack of vowels was a source of confusion. The Greeks solved this problem by repurposing characters for which there was no equivalent sound in the Greek language to represent vowels.

RELATED TOPICS
See also
THE PRINTING PRESS
page 96

THE COMPUTER
page 98

THE TELEGRAPH
page 100

3-SECOND BIOGRAPHY
PIERRE MONTET
1855–1966
French explorer who in 1923 discovered the sarcophagus of King Ahiram with its inscription in Old Phoenician alphabet in Jbeil, Lebanon

30-SECOND TEXT
Diana Rawlinson

3-SECOND SURVEY
The most widely used alphabets of the modern world include Greek, Cyrillic, Arabic and Latin, all of which have an ancient common ancestor in the Near East.

3-MINUTE OVERVIEW
The creation of the alphabet was prompted by the needs of trade and exploration among widely disparate peoples. Commerce, diplomacy and even war required the accurate recording and transmission of information. The simplification achieved by the adoption of consonants with a reduced number of abstract symbols, and the subsequent inclusion of vowels, increased accessibility and provided an environment that allowed for the development of codified law, logic, science and history.

Any alphabet provides the means to write one or more languages using letters. Some early abudgida (alphabets without vowels) are still in evidence today in Arabic and Hebrew.

THE PRINTING PRESS

the 30-second history

Johannes Gutenberg was the first person to demonstrate the practicability and usefulness of a 'movable type machine' around 1440 in Mainz, Germany. A goldsmith by trade, he was quick to recognize the market potential of the mass production of books as literacy among the lay people of Europe spread during the Renaissance. Gutenberg's innovation combined individual letters uniform in size and cast from a metal alloy of lead, tin and antimony, a modified screw press, and oil-based inks to produce a durable and reusable technology. Each individual letter of the alphabet could be moved around to form words, words were separated by blank spaces to form lines of print and those lines were brought together in a galley to make up a page. The capacity to reset the type in any format made it possible to create an infinite variety of texts. The immediate effect of the printing press was to increase the availability of books and make them cheaper to acquire. This in turn drove the dissemination and standardization of information, crucial to the development of education, science and technology. Despite Gutenberg's efforts to keep his techniques secret, more than 250 European cities had printing presses by 1500, producing an estimated 20 million volumes. His most famous printing, some 180 copies of the Bible, sold out at the Frankfurt Book Fair of 1455; remarkably, roughly 50 survive today.

RELATED TOPICS
See also
TIM BERNERS-LEE
page 102

THE SMARTPHONE
page 108

3-SECOND SURVEY
The invention of the printing press introduced books to the masses as well as heralded the demise of Latin as the 'lingua franca'.

3-MINUTE OVERVIEW
Four centuries before Gutenberg, Chinese artisan Pi Sheng was composing texts using characters moulded from a mixture of clay and glue set on an iron plate. Due to the complexity of the Chinese alphabet, which was primarily pictographic rather than alphabetic, and the lack of press technology at the time – the Chinese made neither wine nor olive oil! – printing presses did not take off in China at this time.

3-SECOND BIOGRAPHIES
PI SHENG
990–1051
Chinese artisan who invented the first movable type by carving characters into clay

JOHANNES GUTENBERG
1398–1468
German goldsmith and stonecutter who invented a 'movable type machine'

ALDUS PIUS MANUTIUS
1449–1515
Italian humanist who founded the Aldine Press in Venice in 1494, producing respected editions of the classics

30-SECOND TEXT
Diana Rawlinson

Gutenberg's Bible was the world's first substantial work to be printed, rather than hand copied.

THE COMPUTER

the 30-second history

3-SECOND SURVEY

Computers have enhanced human performance in many fields and permeate modern life to an extent that gives rise to disquiet about our reliance on them.

3-MINUTE OVERVIEW

Computers enable us to carry out many tasks, from sending people to the Moon to sorting through our photographs. But they accomplish this by following pre-set instructions; none has yet passed the 'Turing Test' – they do not pass for human in a comparative test between person and machine. As yet, they don't 'think'. This goes to the heart of the question of what it means to be human.

Before the Industrial Revolution, machines were designed to do one thing – to tell the time or grind corn, for example. But greater mechanization called for something more adaptable. In 1801, Joseph Jacquard built a loom in which interchangeable punch cards were used to control the patterns of cloth produced: the first machine that could be instructed to do more than one task. It was the inspiration for Charles Babbage's 'Difference Engine', conceived in 1822. His idea was for a universal machine capable of limitless calculations that would minimize error and analyse the accuracy of the results. Financial and technological limitations meant that Babbage's machine was never built in his lifetime, but perhaps the reason it took until the twentieth century for the first programmable computers to be developed was that the demand didn't exist. Yet as engineering and military needs became more complex, particularly during the Second World War, issues requiring computations beyond the capabilities of humans in scale and time created that demand. When Alan Turing published his hypothesis of a Turing Machine in 1936 (defined as a machine that can perform any computation that could be performed by any other computing device), he formalized the basis of computing, which together with the application of electronics, provided the foundation for modern computers.

RELATED TOPICS

See also
THE TELEVISION
page 106

THE SMARTPHONE
page 108

3-SECOND BIOGRAPHIES

ADA LOVELACE
1815–52
English mathematician who wrote an algorithm for use by Babbage's analytical engine

KONRAD ZUSE
1910–95
German inventor who built the first fully program-controlled computer, the Z1, in 1938 and the first functional programmable computer, the Z3, in 1941

ALAN TURING
1912–54
English mathematician whose analysis of methodical process led to the development of 'definite method' – in modern language, an algorithm

30-SECOND TEXT
Diana Rawlinson

ENIAC, the world's first electronic computer, cost $500,000 and occupied the same floor space as a small house.

THE TELEGRAPH

the 30-second history

3-SECOND SURVEY

Instantaneous communication through use of the telegraph changed the way people thought about how the world was organized.

3-MINUTE OVERVIEW

The electric telegraph has been called 'the Victorian Internet' because of the similar effects that both inventions had on society and the mixed reception they received. Both use coding to convey messages, creating divisions between those adept at manipulating them and those who only receive the message. Both are subject to concerns about security and notions of information sensitivity, and both transformed connectivity between people, with the potential to unite and to divide.

In the early nineteenth century, scientists were searching for a communication system that could cross large distances without relying on lines of sight. The telegraph emerged when they found they were able to use new developments in the study and understanding of electricity. In Britain, William Fothergill Cooke and Charles Wheatstone produced an electric telegraph that was used for railway signalling, while in the United States, Samuel Morse, Alfred Vail and Leonard Gale developed a battery-powered system needing only a wire to connect to the receiver and a key to tap out the dots and dashes of Morse code. Winning government money to build a telegraph line from Washington, DC, to Baltimore, Morse and Vail sent their first 19-letter message in May 1844. It read: 'What hath God wrought.' The telegraph's potential to speed up communications was soon obvious to commercial interests, which contributed to the rapid expansion of wire connections across the country. The impact was to increase the volume of commerce; markets could be more efficiently exploited and information more easily transferred. This in turn led to more centralized business and administrative systems. The electric telegraph heralded a revolution in communication that highlighted the value of knowledge itself and changed the face of diplomatic and military affairs.

RELATED TOPICS

See also
THE COMPUTER
page 98

THE RADIO
page 104

THE TELEVISION
page 106

3-SECOND BIOGRAPHIES

ALESSANDRO GIUSEPPE VOLTA
1745–1827
Italian physicist and chemist who developed the first battery and discovered methane gas

SAMUEL FINLEY
BREESE MORSE
1791–1872
American portrait painter and inventor of the electric telegraph, who co-invented Morse code with Alfred Vail

30-SECOND TEXT

Diana Rawlinson

The electric telegraph marked the beginning of experimentation with unseen forces, using energy itself.

8 June 1955
Born in London, England to Conway Berners-Lee and Mary Lee Woods

1976
Awarded a first-class degree in physics at the Queen's College, Oxford

1980
While working at CERN, he first proposes the concept of using 'hypertext' to facilitate the sharing of information

1989
Produces 'Information Management: A Proposal' – a system for sharing and distributing information globally; he calls it the World Wide Web

1991
Creates the first web browser and editor, and launches the first ever website, built at CERN 'info.cern.ch'. It explains the World Wide Web and gave users an introduction to getting started with their own websites

1994
Founds the world wide web (W3C) consortium committed to open web standards at the Laboratory of Computer Science at the Massachusetts Institute of Technology

1999
Named one of the 100 Most Important People of the Twentieth Century by *Time* Magazine

2004
Knighted and awarded the Finland Millennium Prize

2007
Awarded the UK's Order of Merit, a personal gift of the monarch, and limited to 24 living individuals.

2009
Elected a fellow of the American Academy of Arts and Sciences

2011
One of the three recipients of the Mikhail Gorbachev award for The Man Who Changed the World

2012
Co-founds the Open Data Institute advocating for open data globally

2012
Plays a starring role in the opening ceremony of the London Olympics viewed by 900 million people; he tweets 'this is for everyone'

TIM BERNERS-LEE

Tim Berners-Lee's fascination with things electronic began when he was a child, creating gadgets to control his model trains. Both his parents were computer scientists and worked on one of the first commercially built computers, the Ferranti Mark I, so his career choice was perhaps no surprise.

After graduating from Oxford with a degree in physics, Berners-Lee worked as a software engineer and consultant, including at CERN, the particle physics laboratory in Geneva. It was here that he first addressed the problem of information sharing with a visionary approach that grew to become the World Wide Web.

The 'Internet', simply meaning the linking of one computer to another, had begun in the 1960s among researchers and academics as a way of transferring content. The limitation that Berners-Lee identified at CERN was that each research team collected their data on a different network of computers, sometimes using a different programming language, which complicated participation and collaboration. His solution was to combine the internet with an emerging technology called hypertext, a system that allowed onscreen viewing of documents from anywhere on the network, without having to download them. His prototype software was called ENQUIRE.

Berners-Lee's 1989 proposal for the World Wide Web was not taken up as an official project at CERN, but he continued to work on it, and by 1990, together with informatics engineer Robert Cailliau, he had written the three fundamental technologies that remain the foundation of the web: HTML (HyperText Markup Language), URL (Uniform Resource Locator) and HTTP (Hypertext Transfer Protocol), and produced the first web browser and the first web server.

From the beginning, Berners-Lee realized that the true potential of the web could only be exploited if the underlying code was available royalty-free so that anyone could use and develop it, and he has continued to advocate for open data, net neutrality and for the web to serve the global public good. Beyond his technical contribution, this commitment to open resources is his most significant, and one that we see spreading beyond the technology sector to politics, scientific research, education and culture.

Diana Rawlinson

THE RADIO

the 30-second history

In 1886 Heinrich Hertz showed that rapid variations of electric current could be projected into space in the form of radio waves. This breakthrough led to the first incarnation of wireless communication as Morse code-based 'wireless telegraphy' patented by Guglielmo Marconi in England in 1896 and simultaneously by Nikola Tesla in the United States. In Calcutta, at the same time, J. C. Bose invented the coherer, allowing radio waves to pass through walls or water. Due to its technical limitations, wireless transmission at this time was used principally for military and particularly maritime endeavours, where it transformed the safety of ocean travel, but had little impact on the rest of society. Further technological improvements were needed before voice broadcasting became possible: in 1906, Reginald Fessenden broadcast a limited programme of speech and music from Brant Rock in Massachusetts. Official concerns about the security of wireless transmission stifled further growth of radio during the First World War but in 1919 wartime controls were relaxed and the development of speech broadcasting began in earnest. People quickly adjusted to 'tuning in' to hear about national and international events as they unfolded, ushering in the era of mass media.

RELATED TOPICS
See also
THE COMPUTER
page 98

THE TELEGRAPH
page 100

THE TELEVISION
page 106

3-SECOND SURVEY
Radio was the first technology to make it possible for people to participate in cultural events in real time and regardless of economic status.

3-MINUTE OVERVIEW
The growth of radio from the 1920s onwards provided unprecedented access to entertainment and information and proved a powerful medium for education. Requiring no special knowledge to operate, the radio allowed people to hear about the latest societal trends. Broadcasting had an immediacy that emphasized interconnectedness and mobilized opinions. The proliferation of wireless sets in the home spawned whole new industries in programme production, advertising and newscasting.

3-SECOND BIOGRAPHIES
NIKOLA TESLA
1856–1943
Serbian-American electrical engineer who patented the alternating current induction motor. Locked in dispute with Marconi over the invention of the wireless, in 1943 Tesla was posthumously awarded the patent in the United States

GUGLIELMO MARCONI
1874–1937
Italian inventor who developed the first long-distance wireless telegraph and in 1901 broadcast the first transatlantic signal

30-SECOND TEXT
Diana Rawlinson

Advances in electronics in the 1900s led to wireless telegraphy becoming an entertainment industry.

THE TELEVISION

the 30-second history

3-SECOND SURVEY
Television entertains and informs us: it provides a sense of connectedness due to its 'live' nature and its popularity has transformed family and social interaction.

3-MINUTE OVERVIEW
In the twenty-first century, many households include more televisions than people, and viewing has had a significant impact on how we relate and influenced the values and norms of society. Television has been blamed for everything from obesity to an increase in violence and terms such as 'boob tube' and 'idiot's lantern' capture the cautionary sense that while it can be an educational medium it does not always broaden the mind.

From 1900, inventors in Britain, Japan, Germany and the United States were experimenting with ideas that would become the television, and dozens of partially successful attempts were produced. In 1925, John Logie Baird amazed London shoppers by showing the first indistinct images, but faced stiff competition in the race to produce a commercially viable system. Within two years, Baird's 'mechanical television' had been superseded by the 'all-electronic' version developed by Vladimir Zworykin and Philo Farnsworth in the United States. Still, use of TV sets was not widespread, and the content limited to barely discernible figures filmed with a single camera. During the Second World War, developments in commercial television were abandoned or even prohibited, but after 1946 they relaunched with an improved user experience, due to technology derived from wartime innovations in radar. Ownership of sets mushroomed; money poured in from advertising and sponsorship, which became an important feature in broadcasting. Television changed what people did with their leisure time and, despite its entertainment and educational benefits, the growth of watching television attracted critics, with the recognition that whoever controlled the content wielded great power, and that its promotion of consumerism could be divisive.

RELATED TOPICS
See also
THE COMPUTER
page 98

THE RADIO
page 104

THE SMARTPHONE
page 108

3-SECOND BIOGRAPHIES
PAUL GOTTLIEB NIPKOW
1860–1940
German engineer who was the first to patent the 'mechanical television' in 1884 using perforated rotating discs; his invention was theoretical

JOHN LOGIE BAIRD
1888–1946
Scottish inventor who demonstrated the first working television, the first colour television tube and in 1928 achieved the first transatlantic transmission

30-SECOND TEXT
Diana Rawlinson

TV is arguably the most widespread invention of the twentieth century, providing a window on the world for every sector of society.

THE SMARTPHONE

the 30-second history

The first phone to combine the functions of a mobile phone with a personal digital assistant was developed by IBM in 1992 and marketed two years later by BellSouth. Named Simon, it was capable of sending and receiving emails and faxes and had features for keeping notes, storing addresses and scheduling appointments. The addition of QWERTY keyboards made it easier to work on the move using word processing and spreadsheets. The Nokia 9000, released in 1996, included a touchscreen with stylus and offered web browsing – unlike Simon. With Penelope, the Ericsson G.S. 88 in 1997, the term smartphone was officially coined. The plethora of similar phones that followed shared impediments that slowed smartphone penetration of the wider consumer market: apps were limited, resistive screens required pressure that was not finger-friendly and web pages were often truncated versions of the PC experience. In 2007, the Apple iPhone broke the mould with capacitive digitizer technology bringing a true touchscreen, and incorporating the iOS WebKit browser (quickly followed by an Android OS) allowing for full-colour, completely rendered web pages. This in turn led to the development of a host of apps for everyday and social functions, transforming the smartphone from a business tool into an object of consumer desire.

3-SECOND SURVEY
The smartphone has quickly become an indispensable conduit to our digital social lives, the irony being that we have less time for face-to-face connections.

3-MINUTE OVERVIEW
Smartphones are practically ubiquitous, allowing us to carry out the administrative aspects of our lives and stay connected wherever we may be. This is a convenience that we have eagerly adopted, but there are drawbacks to always being reachable: it can mean that work never stops; critics talk of overuse leading to addiction, bad posture and anxiety; and new phenomena such as trolling and cyberbullying have become the negative face of constant connectivity.

RELATED TOPICS
See also
THE COMPUTER
page 98

THE TELEVISION
page 106

3-SECOND BIOGRAPHY
THEODURE GEORGE 'TED' PARASKEVAKOS
1937–
Greek-American inventor and electronic engineer who pioneered the transmission of electronic data through telephone lines; he patented the concept of combining data processing and visual-display screens with telephones in 1973

30-SECOND TEXT
Diana Rawlinson

The smartphone puts the power of desktop computing into your pocket – provided you can get a signal – introducing a potential that has yet to be fully exploited.

ECONOMICS & ENERGY

THE ABACUS

the 30-second history

Nearly every ancient culture used something like an abacus to do mathematical calculations – right back to the old Babylonian empire, which used a counting system based on the number 60. No abacuses or counting boards from that era (c. 1900 BCE) have been found, probably because they were made of wood and have rotted away, but images of them do survive. Later the ancient Egyptians, Greeks, Chinese and Romans all used a device like an abacus, moving beads quickly across a board to add or subtract, while the Aztecs in Central America used a similar gadget with knots on a piece of cord to help them with their number system, which was based on 20. There is some evidence that the Roman abacus was copied from the Chinese one; not only were the calculating machines very similar, but there were also trading links between the two countries. The problem in Europe was that Roman numerals were difficult to calculate with. The mathematician Gerbert of Aurillac, who became Pope Sylvester II, reintroduced the abacus into Western Europe, having learned about Arabic numerals from Arab scholars in Cordoba and Seville – though he did so without a numeral for zero. The abacus lasted until the age of the slide rule and pocket calculator.

RELATED TOPIC
See also
PAPER MONEY
page 116

3-SECOND BIOGRAPHIES
GERBERT OF AURILLAC
(POPE SYLVESTER II)
946–1003
French mathematician who introduced the Arabic numeral system to Europe and brought with it the abacus

TERENCE V. ('TIM') CRANMER
1925–2001
American founder of the International Braille Research Center and inventor of the Cranmer Abacus, used to this day by blind people

30-SECOND TEXT
David Boyle

Fast and effective, the abacus led to the development of Arabic numerals.

PAPER MONEY

the 30-second history

Imagine you had spent your whole life, as your ancestors had spent theirs, regarding money as an item that had a value in its own right – and suddenly you were offered a piece of paper standing for money instead. That was what happened to Venetian explorer Marco Polo when he reached the court of Kublai Khan at some point in the 1260s. He was the first European to encounter the phenomenon of paper money, noting with astonishment that it wasn't an option to refuse it. The Great Khan insisted and his insistence carried the force of law. Nobody knows who actually invented paper money. It seems to have evolved from receipts and promissory notes, emerging for the first time during the Chinese Tang dynasty in the seventh century CE, and again when copper was in short supply for coins – which in any case Chinese merchants found too heavy to carry around. The first European banknotes emerged as late as 1661 in Stockholm, as a promise to pay in gold or silver. Yet delinking money from an intrinsic value did carry a huge risk: the inflation and sudden deflations could be terrifying. In fact, a Dutch merchant, Johan Palmstruch, who introduced those first notes in Sweden, was condemned to death for irresponsible book-keeping and died in prison.

3-SECOND SURVEY
Paper money boldly delinks money from the intrinsic value it represents, making it cheap to produce – usually out of paper or plastic.

3-MINUTE OVERVIEW
Paper money separates money from gold, which is generally regarded as a good thing, but it can also – as economist John Maynard Keynes warned – be a weapon of mass destruction. Both the American and French revolutions have been blamed on paper money. In fact, the French aristocracy was turned completely upside down by the boom and bust ushered in by the Scottish economist John Law's issue of banknotes in France in the early eighteenth century.

RELATED TOPICS
See also
E-MONEY
page 128

THE CREDIT CARD
page 130

3-SECOND BIOGRAPHIES
KUBLAI KHAN
1215–94
Ruler of the Mongol empire, one of the first recorded rulers to have specifically ordered paper money

JOHN LAW
1671–1729
Scottish gambler and paper money pioneer who came close to bankrupting France with his experiments

BENJAMIN FRANKLIN
1706–90
American founding father who made the case for paper money to a suspicious British House of Commons

30-SECOND TEXT
David Boyle

China had used paper currency for five centuries before European countries followed suit.

NUCLEAR POWER

the 30-second history

US president Dwight Eisenhower's administration first used the phrase 'atoms for peace' as part of an attempt to make nuclear weapons seem a little more hopeful. It was a reaction against the helter-skelter rush to develop nuclear technology during the Second World War, to beat the Nazis to the key. It had been an eventful story since the physicist Ernest Rutherford first bombarded lithium atoms with protons and found that a huge amount of energy was released. The critical experiments were carried out in Germany in 1938 by Otto Hahn and Fritz Strassmann, with Austrian physicist Lise Meitner and her nephew, Otto Frisch. But within a year, the war had begun and many of the world's best physicists were refugees in the United Kingdom or the United States. The first working reactor went live as an experiment in 1942 and was called Chicago Pile-1; this fed into the Manhattan Project designed to make a nuclear bomb. The first working nuclear reactor to actually produce any electricity was the small Experimental Breeder Reactor in Idaho in 1951. Since then, nuclear energy has struggled to live up to early expectations, mainly because of the sheer expense.

RELATED TOPIC
See also
THE WIND TURBINE
page 124

3-SECOND SURVEY
If splitting the atom could produce energy, it made sense to harness it for the good of the world – but it hasn't always proved easy.

3-MINUTE OVERVIEW
Two nuclear accidents have thrown the development of civil nuclear energy into doubt – Chernobyl in the Ukraine in 1986, where safety systems were shut down, and Fukushima in Japan in 2011, the result of an earthquake and tsunami. In both cases, immediate deaths were relatively low but the land around will not be usable for two generations and the costs of clean-up and precautions at other plants may be immeasurable.

3-SECOND BIOGRAPHIES
ERNEST RUTHERFORD
1871–1937
New Zealand-born physicist whose work led to the first splitting of the atom

OTTO HAHN
1879–1968
German nuclear physics pioneer, who contemplated suicide when he discovered that atomic bombs had been dropped on Hiroshima and Nagasaki

FRITZ STRASSMANN
1902–80
German chemist who helped Otto Hahn discover the phenomenon of nuclear fission

30-SECOND TEXT
David Boyle

In spite of being heralded as 'clean energy', accidents and expense have given nuclear power an uncertain future.

1883
American inventor Charles Fritts creates a solar cell using gold and selenium; it is around 1 per cent efficient

31 January 1898
Russell Ohl born near Allentown, Pennsylvania

23 February 1940
Works on the silicon crystal with a crack down it that leads to his surprise discovery

1946
Gets his patent for solar cells

23 January 1948
Discovers a transistor based on Ohl's N-P barrier

25 April 1954
Bell Laboratories announce the invention of the first practical silicon solar cell

14 May 1973
Launch of the space station Skylab, powered by solar panels

20 March 1987
Dies in California

2008
A new record is broken in solar cell efficiency when the US National Renewable Energy Laboratory creates a cell generating power at more than 40 per cent efficiency

RUSSELL OHL

During his first year at college

Russell Ohl heard his very first radio message. It was an SOS from a ship out in the Atlantic being attacked by a U-boat during the First World War. He had gone to Pennsylvania State University at the age of just 16, and enrolled on the fearsome electrochemical engineering course – at the end of the first year, only three of the original students remained. The radio transmission had caught his imagination and he was determined to go into radio technology.

As a result of that fascination, Ohl stumbled on a peculiarity in silicon that made it the basis for the invention of transistors and also, more directly, of the modern world of solar cells – getting sunlight to generate electricity.

Ohl began work in the Army Signal Corps and progressed through laboratories at Westinghouse and AT&T before he finally began at Bell Laboratories in New Jersey. It was a frustrating career. He was fascinated by the possibilities of semiconductors that control electrical currents, but was constantly being moved from that work to something more urgent, because it was an Air Force contract or something equally vital.

'There were a great many enemies to this work with semiconductors,' he said later. 'You have no idea how many people opposed that. Vacuum tube people said that there is nothing to it and it is all a lot of tommyrot, and that sort of thing.'

The problem was that existing radio tuners didn't work well at high frequencies, so he began to experiment with semiconductor materials like silicon.

In 1940, while working on a silicon sample with a crack down the middle, Ohl found to his surprise that the current flowed between the two sides when it was exposed to light. By a peculiar error, the silicon atoms on one side seemed to have more electrons than those on the other. The gap became known as the 'N-P barrier' (negative-positive). He found he had made a simple device that turned sunlight into electricity.

While there had been solar cells before using selenium, Ohl's cell was far more efficient. There are those who forecast that solar cells will power the next economic revolution.

David Boyle

THE CALCULATOR

the 30-second history

Electronic pocket calculators were the must-have tools of every executive in the 1970s. The invention transformed daily lives by greatly expanding the maths capabilities of everyone from students to business people. It had taken 2,000 years to progress from the abacus, the ancient world's hand-operated 'calculating board' using beads on rods, to mechanical calculators. Wilhelm Schickard invented the 'calculating clock' in 1623, a mechanical adding/subtracting calculator operated by a series of gears. A device invented by the philosopher Blaise Pascal in 1642 also used geared wheels and could add and subtract two numbers directly and multiply and divide by repetition. In the following century Charles Babbage produced what he called a difference engine, the forerunner of the computer. Mechanical calculators, such as the first commercially viable counting machine – the Arithmometer (manufactured until 1915) – and the US-patented push-button Comptometer, dominated office life up until the 1960s. Yet in barely a decade the calculator was transformed; a bulky, heavy desktop machine that ran on mains power, and cost more than a family car, became a cheap and compact battery-powered device that would slip into a pocket. Innovations such as integrated microchips and electronics that could run on batteries made this transition possible.

RELATED TOPICS
See also
THE COMPUTER
page 98

THE ABACUS
page 116

3-SECOND SURVEY
Abacus, slide rule and then calculating machine – the world seems to have beaten a path towards the pocket calculator, until it was driven out by computers.

3-MINUTE OVERVIEW
When Clive Sinclair introduced his Sinclair Executive in 1972, it was the first truly pocket-size calculator. They sold for £80 each. By the end of the decade, similar machines cost just £5 and the calculator industry was about to be blown away by the development of personal computers.

3-SECOND BIOGRAPHIES
WILHELM SCHICKARD
1592–1635
German-Hebrew professor and astronomer credited with being the creator of the first calculator

BLAISE PASCAL
1623–62
French philosopher who developed the first working calculating machine

CLIVE SINCLAIR
1940–
British inventor and pioneer of the first pocket calculator

30-SECOND TEXT
David Boyle

The pocket calculator evolved from the abacus over four thousand years but technological advance is hastening its demise.

THE WIND TURBINE

the 30-second history

Windmills were the pioneering technology of the twelfth century in Europe, harnessing the power of the wind to mechanically turn grinding machines to make flour and drive small machines. The idea that they might also generate energy had to wait – needless to say – until the invention of electricity. After that, it was the Scots who pioneered the idea, perhaps not surprisingly given the prodigious amount of wind that blows across Scotland. It was a Scottish teacher, James Blyth, who set up the first battery-charging wind turbine to power his holiday cottage in Maykirk, Kincardineshire, in 1887. Wind generators made of steel were a common sight on off-grid farms and remote places in the 1930s, particularly in the United States. There were also efforts to plug wind energy into the grid in places such as Yalta in the 1930s, Vermont in the 1940s, and the Orkneys in the 1950s. But fossil fuels continued to dominate until demand for renewable energy sources put wind energy back on the map, starting in the 1970s. Most modern commercial wind turbines are based on a Danish model, with three light-weight horizontal blades that can swivel to meet the wind at the best angle for harvesting this energy. A generator inside the main turbine body takes kinetic energy from the spinning drive shaft (to which the blades are attached) and turns it into electrical energy.

RELATED TOPIC
See also
NUCLEAR POWER
page 118

3-SECOND BIOGRAPHIES
JAMES BLYTH
1839–1906
Scottish maths teacher who first generated energy from the wind

PALMER COSSLETT PUTNAM
1900–84
American geologist responsible for the first American wind turbine, built in Vermont in 1941, which broke down after 1,000 hours

30-SECOND TEXT
David Boyle

3-SECOND SURVEY
In windy places, it makes sense to find a technology to make sure that the energy in the wind is put to good use.

3-MINUTE OVERVIEW
Wind power cannot be enough on its own because many places have unreliable wind. In some places it blows too fiercely and the windmills may topple. Sometimes the wind doesn't blow at all. So wind power, important as it is – and it can boost the energy independence of communities and nations – does need to be balanced by other sources of energy, too.

Offshore wind farms, most recently built on floating rigs, harness higher offshore wind speeds and attract less opposition than onshore construction.

THE BARCODE

the 30-second history

A conversation overheard between a supermarket executive and a senior lecturer at the Drexel Institute of Technology in Philadelphia in 1948 led student Bernard Silver to pursue the idea of barcodes. The question was about how to develop a system that could identify a product automatically. His first attempt at an answer – with his friend and fellow-student Norman Joseph Woodland – involved ultraviolet ink, which proved rather too expensive. The two friends had their barcode system patented in 1952, by which time Woodland was working to interest IBM in the idea. Ten years later, they sold the rights for just $15,000. It was only when the American electronics giant RCA bought the patent and began to develop a similar system that IBM started to catch up. The first grocery item to be scanned was a bar of chewing gum in Troy, Ohio, in 1974. The gum and the receipt are now on display in the Smithsonian Institution in Washington, DC, and barcodes are, of course, absolutely ubiquitous – on books, cars, letters, packages and passports. They have become a lampooned and central part of modern life. Neither of the inventors made any more money out of it. Silver died of leukaemia in 1963 and Woodland watched bemused as their creation took off across the world.

RELATED TOPIC
See also
THE COMPUTER
page 126

3-SECOND BIOGRAPHIES
NORMAN JOSEPH
WOODLAND
1921–2012
American lecturer in mechanical engineering who first developed barcodes by drawing a series of lines in the sand

BERNARD SILVER
1924–63
American electrical engineer who came up with the original idea for the barcode

30-SECOND TEXT
David Boyle

3-SECOND SURVEY
The barcode – a simple pattern of black lines of varying thicknesses that can be scanned electronically – provides the rapid solution to identifying one item among millions.

3-MINUTE OVERVIEW
Woodland and Silver were not the only people working on the idea. Unknown to them, an executive from Pennsylvania Railroad was trying to work out how you could identify carriages or trucks instantly. David Collins tried a system of blue and red lights called KarTrak. The idea did not take off there but was adopted instead by a New Jersey toll road that was trying to work out which cars had paid their subscriptions.

The simple barcode helps us keep track of everything from travel tickets to shopping.

E-MONEY

the 30-second history

When banks started creating money in the form of loans with a stroke of a pen in a balance sheet – many centuries before the advent of information technology – this went some way to making money the virtual idea it is today. But in those days cash was still represented by a thing – a note or a coin. A cryptologist from Los Angeles called David Chaum is credited with having invented the first digital money, which he imagined in 1981 and then developed and launched under the title DigiCash. It had built-in coding to protect the money and to make it untraceable – most e-money is not these days. Chaum moved on when Digicash unravelled in 1998, and used his ideas instead for voting systems, where similar problems of protecting personal identity needed to be solved. He was followed not far behind by two innovators from the National Westminster Bank in London, Tim Jones and Graham Higgins, who invented the first electronic purse, using – among others – a digital money called Mondex. Mondex was sold to Mastercard but it failed to spread as rapidly as expected. As the digital money pioneers put it: there was a need for their inventions, but not nearly enough demand.

RELATED TOPICS

See also
PAPER MONEY
page 116

THE CREDIT CARD
page 130

3-SECOND SURVEY
Cash can be heavy to carry, dirty to count and expensive to guard; it may be more convenient to use blips on a screen.

3-MINUTE OVERVIEW
E-money allows you to exchange small sums with a click. Pioneers include Michael Linton in Canada, with his Local Exchange and Trading Systems (1983) and the mysterious and unidentified Satoshi Nakamoto, who launched bitcoin (2008), using the so-called 'blockchain' technology that encrypts the ledger in every transaction.

3-SECOND BIOGRAPHY
DAVID CHAUM
1955–
American cryptologist and mathematician who launched the first digital money, DigiCash

30-SECOND TEXT
David Boyle

Ethereal perhaps but e-money needs to be as secure as a bank vault.

THE CREDIT CARD

the 30-second history

It all began with a lost cheque book. Financier Frank McNamara was having lunch with his friend Alfred Bloomingdale at Major's Cabin Grill restaurant, next to the Empire State Building in New York City, when they found neither of them had enough cash to pay the bill. The story goes that McNamara had to phone his wife and ask her to drive round with some money. But it was also a revolutionary moment. The two men spent the rest of the meal discussing an idea that could prevent that kind of inconvenience arising again. The following year, in 1950, McNamara launched the Diners' Club, the world's first credit card. But the real shift happened eight years later when a Bank of America middle manager, Joe Williams, organized the famous Fresno Drop, which involved sending 60,000 credit cards out to the inhabitants of Fresno, California, whether they had asked for one or not. It wasn't just the beginning of credit cards in the mass market – it was also the beginning of easy credit for the masses in the United States, which kick-started the consumer revolution. The Diners' Club was an immediate success but McNamara sold his interest a couple of years later and his business is said to have lost a great deal of money in the property market.

3-MINUTE OVERVIEW
If McNamara produced the first credit card, the idea emerged much earlier in an 1888 futuristic novel by American writer Edward Bellamy, called *Looking Backward 2000–1887*. He imagined that everyone would get a money card on which the government would load their monthly allowance. The book was highly controversial and the British artist William Morris was so horrified that he wrote his own version (*News from Nowhere*) without credit cards.

RELATED TOPIC
See also
E-MONEY
page 128

3-SECOND BIOGRAPHIES
EDWARD BELLAMY
1850–98
American writer who first imagined credit cards in his novel about the year 2000

FRANK MCNAMARA
1917–57
American financier who launched Diners' Club and the whole idea of credit cards

30-SECOND TEXT
David Boyle

The idea for the credit card began in a restaurant and now almost everyone in the developed world uses one.

DAILY LIFE

amphorae Large, mostly earthenware, handled containers used for the storage and transport of drink and foodstuff, although typically wine and olive oil. Used for thousands of years from the Neolithic period onwards, they had an airtight seal for preservation; often they had pointed bases, which meant they could be stood upright in soft ground or stored several layers deep aboard trading ships.

arc lamps Extremely bright lights first made in the early 1800s by passing an electric current in air between two electrodes made from carbon. Widely used for lighting streets and factory buildings until replaced by incandescent lights, they remained in use longer for uses such as cinema projection.

bitumen Sticky, black, viscous liquid that occurs naturally as a mix of hydrocarbon substances and is manufactured artificially as a by-product of petrol distillation.

camera obscura Device used for projecting an external view into a closed, darkened space. A camera obscura can range in size from a room to a small box – and works by allowing light to pass through a small hole, in the process inverting the image.

Cinématographe Early example of a convergence device that performed several different functions and was the first mechanism to provide a modern cinema experience in which multiple people could simultaneously watch projected, moving images. It combined camera, printer and projector in one unit.

cooling by vapour compression Process of cooling used in refrigerators and air-conditioning systems. It uses a contained, liquid refrigerant to absorb and remove heat from a given space through a cycle of vaporization, compression and expansion. The same process works whether inside a fridge or a room in a building.

enzymes Biological catalysts that speed up chemical reactions. They each operate best at a particular temperature and in an environment with a particular acidity, or pH level, and are important in the process of respiration. Enzymes are proteins in nature and accelerate reactions by offering a pathway that requires less energy for the reaction to work.

fluorescent bulbs Bulbs that work by passing electric current through mercury vapour (a toxic substance), creating short-wave ultraviolet light that in turn causes a phosphor coating inside the bulb to glow. They make light in a more energy-efficient manner than incandescent bulbs.

Kinetograph First commercial movie camera – a mechanism for synchronizing the uniform movement and exposure of photographic film invented by William Kennedy Dickson.

Kinetoscope Solo mechanism for the projection and viewing of film taken with the Kinetograph.

LED bulbs Bulbs in which electricity is passed through a semiconductor that then emits visible light using a property called electroluminescence – whereby phosphorescent substances emit light particles in response to weak electrical current. LED bulbs produce light more efficiently again than both fluorescent and incandescent light bulbs.

persistence of vision Ability of the brain to retain an image for slightly longer than the time for which it is seen.

phi phenomenon Ability of the brain to edit together a sequence of still images and create the illusion of continuous movement. It does this by filling-in the gaps of information missing between them.

Voltaic pile First electric battery. It was made by piling copper and zinc discs on top of each other, separated by cloth soaked in a salt or 'saline' solution. This worked as an electrolyte – a substance that in solution can conduct electricity – created by differences between the metals.

zero-carbon houses Dwellings that in their construction and operation result in no net carbon emissions being released into the atmosphere. This means, in reality, that their materials must be extremely low-carbon and they must be extremely energy-efficient – utilizing passive solar energy and other renewable sources. To be truly zero carbon, however, some method for storing or 'sequestering' carbon is likely to be needed.

zoetrope A spinning cylinder with uniform viewing slots cut into it that contain static images which when viewed through the slots, as the cylinder spins, create the illusion of moving images.

THE LAVATORY

the 30-second history

Globally, 2.4 billion people lack
access to sanitation – according to the World
Health Organization. Of those, more than
940 million defecate in the open. It's a shocking
statistic for everyone who takes having a toilet
for granted. Domestic toilets connected to
some form of drainage or cesspit are almost
5,000 years old. Public toilets integrated into
urban sewers were important meeting places in
ancient Rome. But the flushing toilet is a more
recent invention. The first working version is
thought to have been designed by English
courtier Sir John Harington in 1596 for his
godmother Queen Elizabeth I. But the flushing
mechanism at the heart of the invention was
invented much earlier as part of a hand-washing
automaton by the Muslim polymath and
mechanical engineer Ismail al-Jazari, who
described his invention in 1206. Flushing toilets
wouldn't reach a broad population, however, until
the Industrial Revolution. A patent for one
incorporating the crucial S-bend was filed by
Alexander Cumming in 1775. Another inventor and
contemporary of Cumming's, Joseph Bramah,
oversaw improvements to the design, which had
a tendency to freeze in cold weather. History
best remembers, however, the man who in the
nineteenth century improved the flushing toilet's
mechanism and popularized it: Thomas Crapper.

RELATED TOPIC
See also
ANTIBIOTICS
page 76

3-SECOND BIOGRAPHIES
ISMAIL AL-JAZARI
1136–1206
Turkish polymath, inventor of
a hand-washing automaton

JOHN HARINGTON
1561–1612
English courtier, author and
inventor of an early flushing
toilet

30-SECOND TEXT
Andrew Simms

3-SECOND SURVEY
Two-thirds of the world's
population can flush away
their waste problem, but
for the rest the door to a
solution remains locked
and 'engaged'.

3-MINUTE OVERVIEW
The United Nations'
Sustainable Development
Goal Number Six is to
'ensure access to water
and sanitation for all'.
Achieving universal
coverage has been costed
at just over $500 billion,
or less than the annual
amount spent on its
military by the United
States. Some charities, like
the Bill & Melinda Gates
Foundation, invest in
high-tech answers to the
world's toilet problem;
for others, the most
sustainable solution is the
humble compost toilet.

*In 1596, Queen
Elizabeth I became
the first monarch to
'sit on the throne' in
the modern sense.*

CANNING

the 30-second history

A 3,000-year-old oak barrel of
butter was dug up in 2009 in County Kildare,
Ireland. It looked uncannily like a rusty food can.
Packaging food for preservation is something
people have done for millennia. The Romans
preserved fish pickle sauce, olive oil, wine and
other foods in sealed amphorae, pipes and
wells. Heating food before placing it in airtight
containers preserves it by killing microorganisms
and 'switching off' enzymes that would lead to
its decay, creating a new protective shield. But
it wasn't until the 1860s, and Louis Pasteur's
work on microorganisms, that sterilization was
understood. The basic practice of this method
of preservation – so-called 'thermal processing'
– was discovered without realizing the science
at work. In 1795 French confectioner Nicolas
Appert began experimenting with airtight
containers using glass bottles and his discovery
was later taken up by the French navy, which
needed a method of preserving food for long
voyages. English inventor Peter Durand used a
mixture of ceramic and glass in his storage
vessels and, in 1810, patented food containers
made of iron coated with tin. Food in tin cans
was made available in the United States in 1822
by Thomas Kensett and Ezra Daggett. Today
cans are mostly made from steel, with around
100 million used daily in the United States alone.

RELATED TOPICS
See also
STEEL
page 24

LOUIS PASTEUR
page 80

REFRIGERATION
page 148

3-SECOND SURVEY
Tin cans revolutionized the
storage and preservation
of food in the nineteenth
century and have been
popular ever since.

3-MINUTE OVERVIEW
Recycling one tin can saves
the equivalent amount of
energy to that used by a
television in three hours, or
74 per cent of the energy
needed to make the can in
the first place. Globally,
only about 8 per cent of
tin actually gets recycled.
From China to Indonesia,
tin mining and smelting
is heavily polluting the
environment. Lighter and
more environmentally
friendly forms of food
packaging now increasingly
replace tin cans.

3-SECOND BIOGRAPHIES
NICOLAS APPERT
1750–1841
French confectioner and
distiller who experimented
with preserving food in heated,
sealed bottles

PETER DURAND
1766–1822
English merchant and inventor
who was given a royal patent
for preserving food in tin-lined
iron cans in 1810

30-SECOND TEXT
Andrew Simms

*Canning may seem like
a modern solution to
extending the life of
perishable food but
the invention has been
around for much longer
than many realize.*

THE LIGHT BULB

the 30-second history

Ironically the invention of the
incandescent light bulb was more of a slow dawn
than a 'light-bulb moment'. While Thomas Edison
receives most of the credit for key light bulb
patents filed by researchers in his company, these
stand in the afterglow of a number of earlier
inventions. At the turn of the nineteenth century,
Italian physicist and chemist Alessandro Volta
developed a zinc and silver battery – or 'voltaic
pile' – with a glowing copper connecting wire.
His invention prefigured the bright filaments of
Edison's bulbs by 80 years. Extremely bright 'arc
lamps' – connecting Volta's batteries with
charcoal electrodes – were developed by English
inventor Humphry Davy in 1802. But it took many
more decades and advances in engineering
before a viable household bulb was produced.
By 1840 a costly bulb with platinum filament
was available. But it required the right filament
material and the creation of a vacuum to make
an efficient bulb that lasted. After decades of
experimentation, Edison bought patents from
others trying to commercialize light bulbs and
tried filaments made from wood, cotton and
thousands of plant materials. His team filed
a key patent in 1879 and discovered that
carbonized bamboo filaments lasted for more
than 1,000 hours. Tungsten filaments, introduced
around 1910, were used until incandescent bulbs
fell out of fashion.

RELATED TOPICS
See also
THE CAMERA
page 142

MOVING PICTURES
page 146

*Economic self-interest
held back the full
benefits of early
household lighting.*

THE CAMERA
the 30-second history

The basic mechanics of the camera are older than most suppose. The way that light passing through a small hole into a darkened space can project an image of what is outside was known to the ancient Greeks and noted by Aristotle. It is the discovery of the chemical process to fix the image that comes much later. The camera obscura (Latin for 'darkened room'), ranged in size from a whole room to a box. These cameras were used as an entertaining novelty and a drawing aid for artists for centuries before the arrival of modern photography. That began with the experiments in heliography, or 'sun drawing', by the Frenchman Joseph-Nicéphore Niépce. After years of chemical experimentation, in 1826–27, he combined a camera obscura and a metal plate covered with a kind of light-sensitive bitumen to make what is considered the first permanent photograph from life – a view from his workshop. The process was slow and difficult. Photography's subsequent development involved reducing the size of the equipment and enhancing the speed and quality of image reproduction. In 1861 a collaboration between English photography pioneer Thomas Sutton and Scottish physicist James Clerk Maxwell produced the first colour photograph. Three negatives separated red, green and blue light in a way that still underpins chemical and electronic methods for recording images.

RELATED TOPICS
See also
THE TELEVISION
page 106

THE SMARTPHONE
page 108

MOVING PICTURES
page 146

3-SECOND SURVEY
For millennia people knew that light passing through a small hole to a dark space projected an image; chemistry then allowed it to be captured.

3-MINUTE OVERVIEW
In its early days photography could dispel myths about everything – from imagined foreign lands to the reality of war. But it could also create new myths, with its apparently unquestionable accuracy. The camera became a source of wonder in the hands of travellers and a tool of control in the hands of the powerful. Now, with a camera in every smartphone, almost everyone is a photographer.

3-SECOND BIOGRAPHIES
JOSEPH-NICÉPHORE NIÉPCE
1765–1833
French inventor who devised the first method for recording a permanent photographic image

THOMAS SUTTON
1819–75
English inventor who took the first colour photo in 1861, using Maxwell's method

JAMES CLERK MAXWELL
1831–79
Scottish physicist whose work on human perception of colour led to colour photography

30-SECOND TEXT
Andrew Simms

A period of 250 years took us from chemistry in a 'darkened room' to 'selfie' over-exposure.

1897
Father is murdered weeks before her birth

10 January 1898
Born in Schenectady, New York

1901
Family moves to France

1912
Family moves back to New York; she attends the Rayson School and becomes interested in mathematics

1917
Awarded a degree in physics from Bryn Mawr College

1918
Completes a masters degree at the University of Chicago and becomes the first woman scientist employed at General Electric's research laboratories

1919
Publishes her first scientific paper, aged 21, on the use of charcoal in gas masks to filter toxins

1926
First woman awarded a PhD in physics by Cambridge University, studying in Ernest Rutherford's Cavendish laboratory

1933
Develops a 'colour gauge' allowing the measurement of layers of film a single molecule thick

1935
Develops a method for spreading layers of film one molecule in thickness

KATHARINE BURR BLODGETT

The legacy of the inventor

Katharine Burr Blodgett is all around you. You probably spend hours every day staring at it, without even noticing. You see it, or rather you don't, whenever you look through certain windows, camera lenses or windscreens; at shop displays or mobile phones; at pictures hung on walls in an art gallery; and in many other circumstances. Blodgett developed a process to eliminate the distorting glare of the reflections from glass; modern society sees more clearly thanks to what was dubbed her 'invisible glass'.

Born in 1898 in Schenectady, New York, Blodgett never met her father, who had been killed by a burglar shortly before she was born. But his influence lived on – she ended up working for the same company as he had done: General Electric. She was encouraged to take up scientific research by one of her father's company acquaintances, Irving Langmuir, who would later win the Nobel Prize in chemistry (in 1932) and whose assistant she would become after graduating from the University of Chicago in 1918.

Later, in 1924, she returned to studying, at Cambridge University in England, where she worked with Ernest Rutherford and became the first woman to be awarded a PhD in physics by the university. Her earlier work with Langmuir investigated the making of ultra-thin, uniform oily layers on water; subsequently she worked on how to transfer the oily films onto hard surfaces and build up multiple layers, which were eventually called 'Langmuir-Blodgett films'. It was only after she developed the process that a practical use for it emerged. Building up a precisely calibrated number of layers of the film on glass, she discovered, had the effect of radically reducing reflection and glare. Invisible glass wasn't her only contribution; during the Second World War she worked on protective measures against poison gas, the de-icing of aircraft wings and smokescreens for ground troops.

Beyond research she engaged in theatre, astronomy and gardening and had two long-term relationships, at different times, with Gertrude Brown and Elsie Errington. Blodgett was given many awards and honorary degrees in her life, but went unmentioned in a 75th anniversary celebration of General Electric's research department published in the journal *Science* in 1953.

Andrew Simms

MOVING PICTURES

the 30-second history

Just as the camera obscura came before the modern camera, another entertainment device, the zoetrope – a spinning drum with gaps cut in its side – prefigured moving pictures. It's a clue to the fact that the real, less acknowledged inventor of moving pictures is the human brain. The zoetrope allowed quick sight of successive still images that the brain, using persistence of vision (in which an image is retained slightly longer than it is seen) and the phi phenomenon (by which a series of still images is seen as continuous motion) combined, to create the illusion of movement. A horse galloping was the first moving photographic picture, produced in 1877 when Eadweard Muybridge, fascinated by the study of animal movement, took a sequence of pictures of a horse. He mounted them in a zoetrope-type spinning disc: when viewed, these settled a long-held debate, proving that all four of a galloping horse's hooves leave the ground simultaneously. The invention in 1888 of celluloid film and the Kinetograph, a mechanism for synchronizing the movement and exposure of film commissioned by Thomas Edison from technician William Kennedy Laurie Dickson, were the next steps forward but only allowed solo viewing on a Kinetoscope. The technology for projection – the cinématographe – was invented by Auguste and Louis Lumière and launched in 1895.

RELATED TOPICS

See also
THE TELEVISION
page 106

THE SMARTPHONE
page 108

THE CAMERA
page 1442

3-SECOND SURVEY
Mechanical improvements allowing rapid sequence photography tapped into the brain's ability to hold and edit images together, creating the illusion of movement.

3-MINUTE OVERVIEW
Moving images allowed people both to reflect on and to escape reality – perhaps the secret of their immense appeal. Some of the earliest films were simple sequences of crowds, street and domestic scenes. But the novelty of appearing to capture familiar faces and places in time, validating the viewers' daily experience and position in the world, was followed quickly by fantasy work like *A Trip to the Moon* by French filmmaker Georges Méliès in 1902.

3-SECOND BIOGRAPHIES
EADWEARD MUYBRIDGE
1830–1904
English photographer whose studies of animal motion created the first photographic moving images

ALICE GUY-BLACHÉ
1873–1968
Early French film pioneer who experimented with sound, colour, special effects and multiracial casting, and maker of possibly the first narrative film, *The Cabbage Fairy*, in 1896

30-SECOND TEXT
Andrew Simms

When reality is either too much or not enough, moving pictures take us somewhere else.

REFRIGERATION

the 30-second history

3-SECOND SURVEY
A Scottish physician's curiosity about the ability of gases and partial vacuums to remove heat led toward refrigeration.

3-MINUTE OVERVIEW
Artificial refrigeration makes life much more comfortable, but comes at a huge environmental cost. Impacts range from waste built up by refrigerator disposal, to the amount of energy demanded and the heavily polluting chemicals used. When it was discovered that the chemicals used in modern refrigeration attacked our protective ozone layer, new ones were found. But these turned out to cause global warming. Cool ways to keep cool are still being sought.

Heat accelerates decay, therefore finding ways to keep things cold has been a priority in human history. Before modern refrigeration there were Victorian ice houses; Arctic Inuit people had little problem preserving food for long periods using ice. Domestic refrigerators are a twentieth-century invention, but artificial cooling using a process called 'vapour compression' to remove heat from a confined space can be traced to demonstrations by Scottish physician William Cullen in the 1740s. It was a century later that commercial cooling systems were being introduced for industry, for example by journalist and mechanical engineer James Harrison, who developed refrigeration for the meat industry, that was patented in 1856 and in use soon after. Today's domestic fridges meanwhile emerged in the nation that is still today synonymous with household appliances: the United States. Fred W. Wolf, son of an industrial refrigeration pioneer in Fort Wayne, Indiana, introduced a commercial electric household fridge in 1913. A year later in Detroit, Michigan, Nathaniel B. Wales developed a unit he named the 'Kelvinator' after physicist William Thomson (Lord Kelvin), who discovered absolute zero. With many similar products coming onto the market by then it is impossible to say that the refrigerator had a single inventor.

RELATED TOPIC
See also
CANNING
page 138

3-SECOND BIOGRAPHIES
WILLIAM CULLEN
1710–90
Scottish physician and celebrated teacher, author of the essay 'Of the Cold Produced by Evaporating Fluids and of Some Other Means of Producing Cold'

FRED W. WOLF, JR
1879–1954
American inventor, first to develop and mass market a household fridge, known as the DOMELRE, short for Domestic Electric Refrigerator

30-SECOND TEXT
Andrew Simms

Refrigeration made food last longer, but wasn't an environmental free lunch.

CENTRAL HEATING

the 30-second history

One reason for human dominance on Earth is our ability to survive in a wide range of environments, from hunting on permafrost to trekking in deserts. But to really thrive we had to learn to control the temperature of our homes. Before advanced engineering, this meant clever building to suit the environment – thick walls for insulation and holes in the right places for ventilation. Then came central heating; it's older than you think. Romans used underfloor and wall heating, blasting hot air from furnaces through clay pipes, a trick some ancient Greeks used, too. However, demonstrating that progress is never a straight line, the idea of central heating was largely lost for centuries. Leap forward to the heart of the Industrial Revolution in Derby, England, for its recognizably modern reinvention. In 1793, mill owner William Strutt developed a hot air furnace to heat his factory, which was then adapted to improve conditions in hospitals. The Scottish inventor and steam pioneer James Watt developed a working household system of steam heating that was also used in greenhouses. But the award for most iconic invention of central heating – the hot-water radiator – goes to Franz San Galli, a Russian inventor and businessman. His 1857 radiator patent fostered our enduring modern expectation of constant, centrally heated comfort.

3-SECOND SURVEY
For millennia, civilization's elite knew how to keep warm, but modern central heating got us all hooked on constant comfort.

3-MINUTE OVERVIEW
The comfort of central heating comes at a cost. Like the energy for the Industrial Revolution, all the warm air, steam and hot water used to control the temperature of our homes has been produced by burning fossil fuels. Keeping our homes warm and at constant temperatures has ironically fuelled the rapid warming and destabilization of our climate. Now the future of home comfort lies with 'zero-carbon houses' warmed by renewable and passive solar energy.

RELATED TOPICS
See also
THE STEAM ENGINE
page 46

THE INTERNAL COMBUSTION ENGINE
page 48

THE LAVATORY
page 136

3-SECOND BIOGRAPHIES
JAMES WATT
1736–1819
Scottish inventor who applied his knowledge of thermal energy to warming homes

WILLIAM STRUTT
1756–1830
English mill owner who made buildings safer using iron frames, and warmed them with his new heating methods

FRANZ SAN GALLI
1824–1908
Russian inventor who gained the tsar's seal of approval to promote his inventions

30-SECOND TEXT
Andrew Simms

Our search for better balance with the biosphere begins at home.

APPENDICES

RESOURCES

BOOKS

Blessed Days of Anaesthesia: How Anaesthetics Changed the World
Stephanie Snow
(Oxford University Press, 2008)

Computer: A History of the Information Machine
Martin Campbell-Kelly
(Westview Press, 2013)

ENIAC: The Triumphs and Tragedies of the World's First Computer
Scott McCartney
(Walker, 1999)

Guglielmo Marconi: Building the Wireless Age
Tim Wander
(New Generation Publishing, 2015)

History, Theory and Practice of the Electric Telegraph
George Bartlett
(Adamant Media Corporation, 2004)

Johannes Gutenberg and the Printing Press
Diana Childress
(Twenty-First Century Books, 2008)

Philo T. Farnsworth: The Father of Television
Donald Godfrey
(The University of Utah Press 2001)

Rebels on the Air: An Alternative History of Radio in America
Jesse Walker
(New York University Press, 2001)

Röntgen Centennial: X-rays in Natural and Life Sciences
Gottfried Landwehr
(World Scientific, 1997)

Smartphones and Beyond: Lesson From the Remarkable Rise and Fall of Symbian
David Wood
(Amazon Media, 2016 Kindle Edition)

Television: An International History of the Formative Years
R. W. Burns
(Institution of Engineering and Technology 1997)

The Alphabet
David Sacks
(Hutchinson, 2003)

The Gutenberg Revolution: How Printing Changed the Course of History
John Man
(Transworld Publishers, 2010)

The Pill: A Biography of the Drug That Changed the World
Bernard Asbell
(Random House, 1995)

The Smartphone: Anatomy of an Industry
Elizabeth Woyke
(The New Press, 2014)

The Story of Writing: Alphabets,
Hieroglyphs and Pictograms
Andrew Robinson
(Thames & Hudson, 2007)

The Victorian Internet
Tom Standage
(W&N; New Ed Edition, 1999)

Wheels of Fortune – A Salute to Pioneers
Sir Arthur Du Cros
(Chapman & Hall, 1938)

JOURNAL ARTICLES

'A Brief History of Cardiac Pacing'
Oscar Aquilina
(*Images Paediatr Cardiol.*, 2006, 27: 17–81.)

WEBSITES

ancient.eu
Jan van der Crabben, Alphabet
www.ancient.eu/alphabet

engineersgarage.com
Samidha Verma, Invention Story of Radio
www.engineersgarage.com/invention-
stories/radio-history

history-of-the-microscope.org
David Bardell, The Invention of the Microscope
www.history-of-the-microscope.org

thoughtco.com
Mary Bellis, The History of the
Telegraph and Telegraphy
www.thoughtco.com/the-history-of-the-
electric-telegraph-and-telegraphy-1992542

Mary Bellis, The History of Television
www.thoughtco.com/television-history-1992530

NASA (National Aeronautics and
Space Administration)
Global Positioning System (GPS)
www.nasa.gov/directorates/heo/scan/
communications/policy/GPS_History.html

psprint.com
PsPrint, The Invention of the Printing Press
www.psprint.com/resources/printing-press/

turing.co.uk
Andrew Hodges, Alan Turing: The Enigma
www.turing.org.uk

w3.org
World Wide Web Consortium, W3C
www.w3.org

webfoundation.org
World Wide Web Foundation,
The Web Belongs to All of Us
www.webfoundation.org

NOTES ON CONTRIBUTORS

EDITOR

David Boyle is the author of several books about history, the history of ideas and the future. His *Authenticity: Brands, Fakes, Spin and the Lust for Real Life* helped put the search for authenticity on the agenda as a social phenomenon, while *Funny Money: In Search of Alternative Cash* launched the 'time banks movement' in the UK. His work on the history and future of money has also been covered in books and pamphlets such as *Why London Needs Its own Currency, Virtual Currencies, The Money Changers: Currency Reform from Aristotle to E-cash, The Little Money Book* and *Money Matters*. He has stood for Parliament, undertaken an independent review for the UK Treasury and Cabinet Office and written a number of well-received history books, including *Blondel's Song: The Imprisonment and Ransom of Richard the Lionheart, Toward the Setting Sun: Columbus, Cabot, Vespucci and the Race for America, Alan Turing: Unlocking the Enigma* and *Before Enigma: The Room 40 Codebreakers of the First World War.*

CONTRIBUTORS

Judith Hodge has worked as a writer and editor in the UK and New Zealand. She is the author of several books and articles on science, health and green consumerism, including the *New Zealand Green Guide*. After retraining as a public health nutritionist, she is currently co-editor of *Nutrition Exchange*, a publication about nutrition in the developing world.

Diana Rawlinson was the project administrator for both phases 1 and 2 of the DANTE project for Lexicography Masterclass Ltd., creating the *New English–Irish Dictionary*. In 2009–13 she was the legal entity appointed representative for Lexical Computing Ltd on EU Lifelong Learning projects. In 2016 she was a contributor to *The Oxford Handbook of Lexicography*. She now works as a freelance writer.

Andrew Simms is a political economist, environmentalist and co-founder of the New Weather Institute. He is a research associate at the Centre for Global Political Economy at the University of Sussex and a fellow of the New Economics Foundation (NEF), where he was policy director for more than a decade. During that time he founded the NEF's work programme on climate change, energy and interdependence, instigated their 'Great Transition' project and ran work on local economies, coining the term 'clone towns' to describe the homogenization of high streets by chain stores. He wrote the books *Tescopoly* (on Tesco's dominance of the grocery market) and *Ecological Debt* (on framing the transgression of planetary boundaries) and co-authored *The New Economics* and *Green New Deal*, devising the concept of 'ecological debt day' to illustrate when in the year we begin living beyond our environmental means. He writes often for *The Guardian* and was described by *New Scientist* magazine as 'a master at joined-up progressive thinking'. His latest book *Cancel the Apocalypse: The New Path to Prosperity* is a manifesto of new economic possibilities. You can follow him on Twitter at @andrewsimms.uk

INDEX

ACKNOWLEDGEMENTS

The publisher would like to thank the following for permission to reproduce copyright material:

All images from Shutterstock, Inc. and Clipart Images unless stated

Alamy /Blend Images 87BL, /Paul Quayle 109B

Europeana Collections /British Library 9B, 9C, 9T, 15BL, 15BR, 15C, 17TR, 19TL, /Wellcome Library, London 19CL; 23C, /Wellcome Library, London 34CR; 39BL, 42C, 42BL, 42BR, /British Library 47TR; 55T, 55CR, 55BC, /British Library 57TL, /Wellcome Library, London 75, /British Library 77BL; 77CR, 77TR, 79BL, 79TL, 83CL, 85B, 85C, 85T, 95B, 95C, 95CL, 97TC, 99TC, 115TR, 117BC, 117C, 117CR, 117TL, 119C, 123TC, /British Library 125CR; 137BL, 137C, 139TR

Flickr/Ur Cameras 65BR

Library of Congress 17TL, 27C, 29CL, 37, 47TC, 47CL, 49, 57C, 57BL, 60, 63TL, 63TR, 67B, 73CL, 97TR, 97B, 101BL, 101CR, 101TL, 101TR, 105BL, 105CR, 105TC, 117CL, 117TL, 131C, 139BR, 141CT, 149CL, 149CR, 149TR

NASA 67B

National Portrait Gallery, London 40

New York Public Library Digital Collection 65TR, 65TL, 137TL, 137TR, 137BR

Courtesy of the US Naval Research Laboratory 67TC, 157

Okänd/Tekniska Museet 143C, 147C, 147TR

Sarah Skeate 120

Smithsonian Institution Archives 144
Wellcome Library, London 9, 35C, 43, 47

Wikipedia 9C, 15L, 15C, 17C, 17CL, 17TL, 17TR, 19C, 19BR, /Chemical Heritage Foundation 20; 23B, 23CR, 25TL, 25TR, 25C, 25BR, 27TL, 35TL, 39TR, 45C, 45CR, 47CL, /Wellcome Library, London 47TL; 47TR, 59T, 59CL, 59BR, 63C, 63BC, /US Coast Guard 65CR; 79TL, 79TR, 79BR, /Félix Nadar 80; 83TL, 83CT, 85C, 89BL, 89CL, 89C, 97B, /Paul Clarke 102; 107BR, 107CL, 107CR, 107TC, 115CT, 119C, 119CB, 119CL, 119CR, 123BR, 123C, 125BR, 125CL, 127BL, 143CR, 147CL, 147CR, 147TC, 149TR

Yale Center for British Art 23TR, 35BL, 39C